National Aeronautics and Space Administration

GLOBAL REACH
A View of NASA's International Cooperation

To learn more about NASA and the Office of International and Interagency Relations, please visit NASA's Web site at *www.nasa.gov*.

To the Reader:

When the National Aeronautics and Space Administration (NASA) was created in 1958, its founding legislation—the National Aeronautics and Space Act—directed the new Agency to pursue cooperation "with other nations and groups of nations." This principle of international cooperation has been a guiding philosophy for NASA, and it has never been more important than it is today. Such collaboration will be essential to addressing the inherently global and interrelated scientific challenges that will face us in the years ahead: expanding human exploration beyond the frontiers of low-Earth orbit; broadening human knowledge by answering profound questions about the Earth and the universe we live in; solving technical issues related to air traffic management, aviation safety, and the impact of aviation on climate and the environment; and leveraging technology investments to push the boundaries of innovation.

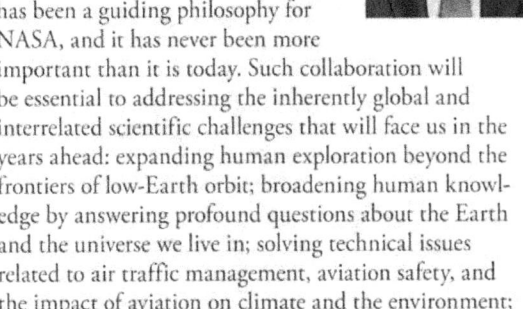

In carrying out its mandate of research and space exploration, NASA has, since its establishment, entered into more than 3,000 agreements with over 120 nations and international organizations. NASA's global partnerships are represented today by more than 600 active agreements with partner entities around the world. These partnerships offer multiple benefits to NASA and its partners, from accelerating the pace of scientific progress through open access to science mission data to sharing risks and costs while promoting discovery and advancement.

While a few longstanding international partners (such as Japan, Russia, Canada, the European Space Agency, and several European nations) account for a large percentage of NASA's international activities, NASA has expanded its "global reach" in recent years to include cooperative activities with new partners in Africa, Asia, and the Americas. This is in part a reflection of growing global interest in the use of Earth science data for societal benefit—particularly in support of local decision-making about such important topics as agricultural productivity, water management, disaster prediction and response, and vector-borne diseases.

Of the many examples of NASA's international cooperation described in this book, perhaps the best known is the International Space Station. This orbiting research laboratory—built by the United States, Canada, Europe, Japan, and Russia—has been called home by a rotating crew of six from around the world for 13 years of continuous crewed operations. In another well-known mission, several international partners contributed to NASA's Curiosity rover, which made headlines in 2012 with its historic precision landing on Mars to investigate whether that planet has ever had an environment capable of supporting life. NASA is continuing to pursue international partnerships on programs that push the boundaries of human and robotic space exploration, such as the James Webb Space Telescope, the Global Precipitation Measurement mission, the Space Launch System, and the Orion crew module. Our international partners are also an intrinsic part of planning for future programs, such as an innovative asteroid initiative that will pave the way for an eventual human mission to Mars by leveraging ongoing investments in space technology, science, and space exploration.

I would like to invite you, through the pages of this publication, to learn more about NASA's continuing international partnerships. I am confident that these examples will provide you with a sense of the breadth of NASA's cooperation with other nations, as well as the accompanying benefits that such cooperation has always yielded.

Michael F. O'Brien
Associate Administrator
Office of International and Interagency Relations

Table of Contents

International Missions and Projects

Argentina
Aquarius/Scientific Application Satellite-D (Aquarius/SAC-D) 2
Scientific Application Satellite-C (SAC-C) 3

Australia
Deep Space Network (DSN) 4
Aircraft Icing Research 5

Austria
Magnetospheric MultiScale (MMS) 6
Time History of Events and Macroscale Interactions during Substorms (THEMIS) and Acceleration, Reconnection, Turbulence and Electrodynamics of the Moon's Interaction with the Sun (ARTEMIS) 7

Belgium
Juno 8
Solar Probe Plus (SPP) 9

Bermuda
Mobile Tracking Station 10

Brazil
Global Precipitation Measurement (GPM) Feasibility Study 11
Ozone Cooperation 12

Canada
Aircraft Icing Research 13
CloudSat 14
International Space Station (ISS) 15
James Webb Space Telescope (JWST) 16
Mars Science Laboratory (MSL) 17
Origins Spectral Interpretation Resource Identification Security–Regolith Explorer (OSIRIS-REx) 18
Resource Prospector Mission (RPM) 19
Robotic Refueling Mission (RRM) 20
SCISAT-1 21
Soil Moisture Active Passive Mission (SMAP) 22
Terra 23
Time History of Events and Macroscale Interactions during Substorms (THEMIS) and Acceleration, Reconnection, Turbulence and Electrodynamics of the Moon's Interaction with the Sun (ARTEMIS) 24

Czech Republic
Van Allen Probes 25

Denmark
LEGO 26
Mars Exploration Rovers (MER) 27
Nuclear Spectroscopic Telescope Array (NuSTAR) 28

- SCIENCE
- HUMAN EXPLORATION AND OPERATIONS
- AERONAUTICS RESEARCH
- EDUCATION AND OUTREACH
- MULTILATERAL REPRESENTATION AND COORDINATION

Table of Contents

Europe (European Space Agency collaborations with individual European countries)
- Cassini-Huygens 29
- Herschel 30
- Mars Express 31
- Planck 32
- Solar Orbiter 33
- Solar TErrestrial RElations Observatory (STEREO) 34

European Space Agency
- Cluster-II 35
- Euclid 36
- ExoMars 37
- Hubble Space Telescope (HST) 38
- International Space Station (ISS) 39
- ISS-Early Utilization 40
- ISS-Human Research Facility-European Physiology Module (ISS HRF-EPM) 41
- James Webb Space Telescope (JWST) 42
- JUpiter ICy moons Explorer (JUICE) 43
- Laser Interferometer Space Antenna (LISA Pathfinder) 44
- Rosetta 45
- Solar and Heliospheric Observatory (SOHO) 46
- X-ray Multi-Mirror-Newton (XMM-Newton) 47

Finland
- Angry Birds Space 48

France
- Aircraft Icing Research 49
- Airframe Noise and Environmental Noise Mitigation 50
- Cloud-Aerosol LIDAR and Infrared Pathfinder Satellite Observations (CALIPSO) 51
- DEvice for the study of CrItical LIquids and Crystallization–Directional Solidification Insert (DECLIC-DSI) 52
- Fermi Gamma-ray Space Telescope 53
- Interior Exploration using Seismic Investigations, Geodesy and Heat Transport (InSight) 54
- Juno 55
- Magnetospheric MultiScale (MMS) 56
- Mars Atmosphere and Volatile Evolution Mission (MAVEN) 57
- Mars Odyssey 58
- Mars Science Laboratory (MSL) 59
- Ocean Surface Topography Mission (OSTM) 60
- Solar Probe Plus (SPP) 61
- Space Environment Testbed-1 (SET-1) 62
- Surface Water Ocean Topography (SWOT) 63

Table of Contents

Germany
- Chandra X-ray Observatory 64
- Computational Fluid Dynamics Research (CFD) 65
- Dawn Asteroid Rendezvous Mission 66
- Fermi Gamma-ray Space Telescope 67
- Gravity Recovery and Climate Experiment (GRACE) 68
- Gravity Recovery and Climate Experiment Follow-On (GRACE-FO) 69
- Interior Exploration using Seismic Investigations, Geodesy and Heat Transport (InSight) 70
- Mars Exploration Rovers (MER) 71
- Mars Science Laboratory (MSL) 72
- Solar Probe Plus (SPP) 73
- Stratospheric Observatory for Infrared Astronomy (SOFIA) 74
- Time History of Events and Macroscale Interactions during Substorms (THEMIS) and Acceleration, Reconnection, Turbulence and Electrodynamics of the Moon's Interaction with the Sun (ARTEMIS) 75
- Two Wide-angle Imaging Neutral-atom Spectrometers (TWINS) 76

India
- Global Precipitation Measurement (GPM)/Megha-Tropiques 77
- Oceansat-2 78

Italy
- BepiColombo 79
- Dawn Asteroid Rendezvous Mission 80
- Fermi Gamma-ray Space Telescope 81
- Juno 82
- Mars Reconnaissance Orbiter (MRO) 83
- Multi-Purpose Logistics Module (MPLM) 84
- Nuclear Spectroscopic Telescope Array (NuSTAR) 85
- Swift Gamma-ray Burst Explorer 86

Japan
- Astro-H 87
- CALorimetric Electron Telescope (CALET) 88
- Disruption Tolerant Networking (DTN) 89
- Fermi Gamma-ray Space Telescope 90
- Global Precipitation Measurement (GPM) 91
- Hayabusa/Hayabusa-2 92
- Hinode 93
- International Space Station (ISS) 94
- Magnetospheric MultiScale (MMS) 95
- Sonic Boom Research 96
- Suzaku 97

Table of Contents

    Terra 98
    Tropical Rainfall Measuring Mission (TRMM) 99

Republic of Korea
    Advanced Colloids Experiment (ACE) 100
    Heliophysics and Space Physics 101

The Netherlands
    Aura 102
    Flight Deck System Design 103

Norway
    Interface Region Imaging Spectrograph (IRIS) 104
    Ka-band Radio Frequency (RF) Propagation
    Monitoring Station 105
    NASA Isbjørn Facility at SvalSat 106

Russia
    International Space Station (ISS) 107
    Lunar Reconnaissance Orbiter (LRO) 108
    Mars Odyssey 109
    Mars Science Laboratory (MSL) 110
    Wind 111

Spain
    Deep Space Network (DSN) 112
    Mars Science Laboratory (MSL) 113

Sweden
    Fermi Gamma-ray Space Telescope 114
    Magnetospheric MultiScale (MMS) 115

Switzerland
    Interstellar Boundary Explorer (IBEX) 116

Thailand
    Vector-Borne Disease Control Research 117

United Kingdom
    Aura 118
    Hinode 119
    Space Environment Testbed-1 (SET-1) 120
    Swift Gamma-ray Burst Explorer 121

Worldwide Missions, Programs, and Networks

    Aerosol Robotic Network (AERONET) 124
    Alpha Magnetic Spectrometer (AMS) 125
    Astrobiology and Analogs 126
    Balloons 127
    Education 128
    Exhibits 129
    Famine Early Warning System Network (FEWS-NET) 130

Global Exploration Strategy (GES) 131
Global Learning and Observations to Benefit
The Environment (GLOBE) 132
International Astronauts 133
International Space Weather Initiative (ISWI) 134
International Space Station Utilization: Science and Technology
Research and Development 135
SERVIR 136
Sounding Rockets 137
Space Geodesy 138

Multilateral Representation and Coordination

- Committee on Earth Observation Satellites (CEOS) 140
- Committee on Space Research (COSPAR) 140
- Consultative Committee for Space Data Systems (CCSDS) 140
- Global Geodetic Observing System (GGOS) 140
- Group on Earth Observations (GEO) 140
- Interagency Operations Advisory Group (IOAG) 140
- Interagency Space Debris Coordination Committee (IADC) ... 140
- International Astronautical Federation (IAF) 140
- International Charter on Space and Major Disasters 141
- International Committee on GNSS (ICG) 141
- International Forum for Aviation Research (IFAR) 141
- International Living With a Star (ILWS) 141
- International Mars Exploration Working Group (IMEWG) 141
- International Ocean-Colour Coordinating Group (IOCCG) 141
- International Space Education Board (ISEB) 141
- International Space Exploration Coordination Group (ISECG) 142
- International Telecommunications Union (ITU) 142
- UN Committee on the Peaceful Uses of Outer Space
  (COPUOS) 142

Index by NASA Mission and Program Area 143

List of Acronyms 147

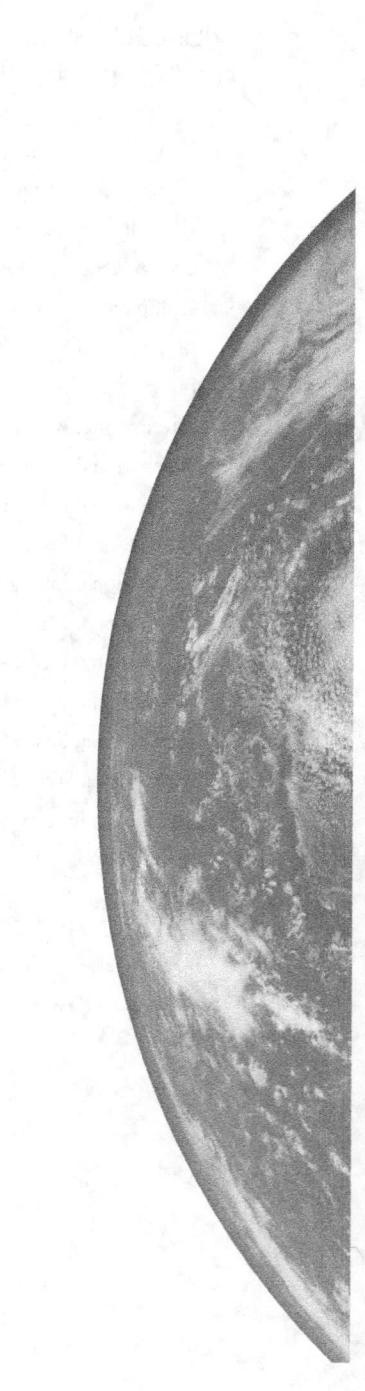

INTERNATIONAL MISSIONS AND PROJECTS

Argentina

Aquarius/SAC-D
Aquarius/Scientific Application Satellite–D

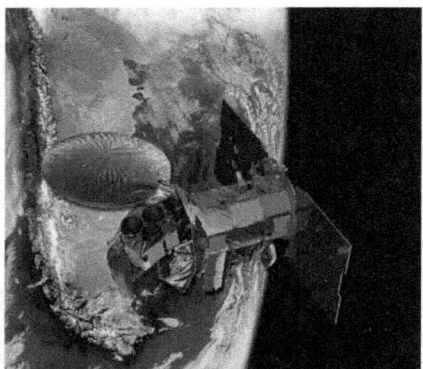

An artist's concept of the Aquarius/SAC-D satellite orbiting Earth. (Credit: Argentine Commission on Space Activities)

NASA and the Argentine National Commission on Space Activities (CONAE) are cooperating on the Aquarius/Scientific Application Satellite–D (SAC-D) mission, which launched in June 2011 from Vandenberg Air Force Base (VAFB), California. NASA provided the Aquarius instrument, which measures global sea surface salinity, and executed the launch of SAC-D. CONAE provided the spacecraft bus, data-collection system, high-sensitivity camera, microwave radiometer, and New Infrared Sensor Technology (NIRST). Through CONAE, France provided the Carmen I instrument and Italy provided the Radio Occultation Sounder for Atmosphere (ROSA) instrument. Data-processing, dissemination, and archiving tasks are performed by NASA. Mission operations are conducted at the CONAE ground station in Córdoba, Argentina.

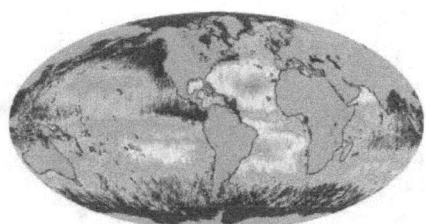

This map shows salinity near the ocean surface as measured by NASA's Aquarius instrument on Argentina's SAC-D satellite. Lower values are represented in purples and blues; higher values are shown in shades of orange and red.

The Aquarius/SAC-D mission provides data that helps scientists better understand the climatic interactions between the global water cycle and ocean circulation by systematically mapping the spatial and temporal variations of sea surface salinity. Aquarius measures sea surface salinity variability—the key tracer for freshwater input and output to the ocean associated with precipitation, evaporation, ice-melting, and river runoff. These measurements, combined with sea surface temperature (SST) from other satellites, assist in determining sea surface density, which controls the formation of water masses and regulate the three-dimensional ocean circulation. The science goals of the mission are to observe and model the processes that relate salinity variations to climatic changes in the global cycling of water in order to understand how these variations influence the general ocean circulation. The Aquarius instrument yielded NASA's first global map of the salinity of the ocean surface, demonstrating Aquarius's ability to detect large-scale salinity distribution features clearly and with sharp contrast.

For more information on the Aquarius/SAC-D satellite, please visit *http://aquarius.nasa.gov/* and *http://www.conae.gov.ar/eng/satelites/sac-d.html*.

Argentina

SAC-C
Scientific Application Satellite–C

The Scientific Application Satellite–C (SAC-C) mission, launched in 2000, is a joint mission between NASA and the Argentine National Commission on Space Activities (CONAE) to study the geomagnetic field, atmospheric structure, and terrestrial and marine environments.

CONAE provided the spacecraft bus, data-collection system, Multispectral Medium Resolution Scanner (MMRS), High-Resolution Panchromatic Camera (HRPC), High-Sensitivity Technological Camera (HSTC), and a whale-tracker experiment. Through CONAE, Italy provided the Italian Navigation Experiment (INES) and Italian Star Tracker (IST), and France supplied the Influence of Space Radiation on Advanced Components (ICARE) payload. NASA executed the launch of SAC-C and also contributed the magnetic mapping payload (MMP) and the Global Positioning System (GPS) Occultation and Passive reflection Experiment (GOLPE).

Although SAC-C lost its maneuvering capability in 2011, many of its instruments continue to provide science data. Argentine and U.S. science teams use data from the SAC-C mission as well as from the Aquarius/Scientific Application Satellite–D (SAC-D) mission, which carries similar instruments on board.

The SAC-C spacecraft is part of the Earth-observing Morning Constellation of satellites, which fly in formation and include SAC-C, Landsat 7, Earth Observing-1 (EO-1), and Terra.

For more information on SAC-C, please visit *http://www.conae.gov.ar/eng/satelites/sac-c.html*.

An artist's concept of the SAC-C satellite orbiting Earth. (Credit: Argentine Commission on Space Activities)

Human Exploration and Operations

Australia

DSN
Deep Space Network

The Australian Commonwealth Scientific and Industrial Research Organization (CSIRO), on behalf of NASA, manages the Canberra Deep Space Communication Complex (CDSCC), located 40 kilometers southwest of Canberra near the Tidbinbilla Nature Reserve. The CDSCC is one of three facilities that comprise NASA's Deep Space Network (DSN). The United States and Australia have engaged in DSN cooperation since the late 1950s, and the CDSCC was officially opened as a NASA DSN facility in 1965.

NASA's DSN is an international network of antennas that communicates with interplanetary spacecraft, is used by radio and radar astronomers to observe the solar system and the universe, and supports Earth-orbiting satellites. NASA's three DSN facilities are Goldstone, in the Mojave Desert, California; CDSCC, near Canberra, Australia; and the Madrid Deep Space Communications Complex (MDSCC), in Robledo de Chavela, Spain.

These DSN facilities are located approximately 120° apart around the world, which permits constant observation of spacecraft as the Earth rotates. Each location has an 8- to 14-hour viewing period for contact with spacecraft. The DSN provides the two-way communications link that tracks, guides, and controls the spacecraft and returns the telemetry and scientific data it collects. Incoming data to the three DSN facilities are processed and transmitted to NASA's Jet Propulsion Laboratory (JPL) in Pasadena, California, for further processing and distribution to science teams over a modern ground communications network.

Each DSN facility consists of three classes of deep space antennas equipped with ultra-sensitive receiving systems and large parabolic dish antennas: a 34-meter-diameter high-efficiency antenna, a 34-meter beam waveguide antenna, and a 70-meter antenna.

NASA is currently building two additional beam waveguide antennas at the CDSCC facility in Canberra to increase NASA's deep space capability in the Southern Hemisphere.

NASA periodically uses CSIRO's 64-meter antenna at Parkes Radio Astronomy Observatory in New South Wales, Australia, to augment CDSCC. In 2002, this antenna was upgraded by NASA to be compatible with future NASA deep space missions.

For more information on the DSN and CDSCC, please visit *http://deepspace.jpl.nasa.gov/dsn/* and *http://www.cdscc.nasa.gov/*.

A view of Canberra's 70-meter (230-foot) antenna with flags from the three Deep Space Network sites.

A view of the Canberra Complex.

Australia

Aircraft Icing Research

The aviation community is tackling a phenomenon called ice crystal engine icing, in which tiny ice particles are ingested by and melt inside warm aircraft engines, forming a thin film of water that covers the engine interior and then traps additional ice crystals. As more ice crystals enter the engine, the melted film of water refreezes, and ice builds up inside the engine. If enough ice builds up and breaks free, it can travel through the engine and damage critical parts, or it may melt and cause the engine to lose power or shut down altogether. Since 1988, safety officials have documented more than 150 such incidents in which jet engines have experienced power loss, affecting various types of jet aircraft from commercial airliners to the smallest business airplanes. In addition to safety concerns, these events can also lead to costly engine repairs.

A computer rendering of the research aircraft NASA will use to collect vital atmospheric data about where, when, and how high-altitude ice crystals occur.

To explore and further the science surrounding ice crystal engine icing, NASA will join a team of European researchers on a flight campaign that will collect data about ice crystals and their environment. A ground research effort will use this data to better simulate aircraft-engine-icing conditions for full-scale engine tests. Researchers hope to answer questions about what causes ice crystals to form, where they are most likely to occur, and under what conditions they become a hazard that leads to engine-icing problems. The primary campaign to collect data will take place in Darwin, Australia.

Polar opposites attract in the puzzling case of ice crystal engine icing, where the frozen crystals can be ingested into the core of a jet engine.

In addition to NASA, the research team consists of scientists from the following governmental and private entities: Airbus, Australia's Bureau of Meteorology, Boeing, Environment Canada, the National Research Council of Canada, Transport Canada, the U.S. Federal Aviation Administration, and the U.S. National Center for Atmospheric Research (NCAR).

For more information on the campaign, please visit *http://www.nasa.gov/centers/glenn/news/pressrel/2010/10-069_water.html* and *http://www.nasa.gov/topics/aeronautics/features/preps_solve_engine_icing.html*.

Austria

MMS
Magnetospheric MultiScale

An artist's concept of the four identical Magnetospheric MultiScale spacecraft investigating magnetic reconnection within Earth's magnetic field. (Credit: Southwest Research Institute)

NASA's Magnetospheric MultiScale (MMS) mission is a solar-terrestrial probe mission comprising four identically instrumented spacecraft that will use Earth's magnetosphere as a laboratory to study the microphysics of three fundamental plasma processes: magnetic reconnection, energetic particle acceleration, and turbulence. These processes occur in all astrophysical plasma systems but can be studied in situ only in our solar system and most efficiently only in Earth's magnetosphere, where they control the dynamics of the geospace environment and play an important role in the processes known as "space weather." The four MMS spacecraft are currently planned for launch by NASA in 2014.

NASA is collaborating on the MMS mission with the Aeronautics and Space Agency (ALR) of the Austrian Research Promotion Agency (FFG) through the Space Research Institute of the Austrian Academy of Science (IWF/ÖAW). NASA is also cooperating with the Japan Aerospace Exploration Agency (JAXA), the French National Center for Space Studies (CNES), and the Swedish National Space Board (SNSB) on the mission.

MMS investigates how the magnetic fields of the Sun and Earth connect and disconnect, explosively transferring energy from one to the other in a process known as magnetic reconnection. This process limits the performance of fusion reactors and is the final governor of geospace weather, which affects modern technological systems such as telecommunications networks, Global Positioning System (GPS) navigation, and electrical power grids. MMS's four spacecraft will measure plasmas, fields, and particles in a near-equatorial orbit that will frequently encounter reconnection in action.

The four MMS spacecraft will carry identical instrument suites of plasma analyzers, energetic particle detectors, magnetometers, and electric field instruments; they will also carry a device to prevent spacecraft charging from interfering with the highly sensitive measurements required in and around the diffusion regions. FFG/ALR is providing the four MMS spacecraft with components for the Active Spacecraft Potential Control (ASPOC) system and the Digital Fluxgate Magnetometer (DFM), as well as gun detector electronics (GDE) for the Electron Drift Instrument (EDI). NASA is providing the spacecraft, the launch, and overall mission management.

For more information on MMS, please visit *http://mms.gsfc.nasa.gov/*.

Austria

THEMIS
Time History of Events and Macroscale Interactions during Substorms
and
ARTEMIS
Acceleration, Reconnection, Turbulence and Electrodynamics of the Moon's Interaction with the Sun

NASA's Time History of Events and Macroscale Interactions during Substorms (THEMIS) mission originated as a 2-year mission consisting of five identical satellites, or probes, launched together in 2007 to study the violent and colorful eruptions in auroras. The mission incorporates a network of ground-based auroral observatories.

An artist's concept of one of the Time History of Events and Macroscale Interactions during Substorms spacecraft orbiting Earth.

THEMIS helps to determine which physical processes in near-Earth space initiate violent substorm eruptions in the Earth's magnetosphere. Substorms intensify auroras and create a dramatic "dancing" effect in them. Aligning five identical probes over observatories on the North American continent has allowed scientists to collect coordinated measurements along the Earth's magnetic field lines, thereby providing the first comprehensive look at the onset of substorms and the manner in which they trigger auroral eruptions.

In 2009, NASA assigned two of the THEMIS satellites to a new mission—the Acceleration, Reconnection, Turbulence and Electrodynamics of the Moon's Interaction with the Sun (ARTEMIS) mission—to measure solar wind turbulence at the Moon. Having repositioned two of the five THEMIS probes in coordinated, lunar-equatorial orbits, ARTEMIS is now performing the first systematic, two-point observations of the distant magnetotail, the solar wind, and the lunar space environment. The primary objectives of the mission are to study how particles are accelerated at reconnection sites and shocks and how turbulence develops and evolves in Earth's magnetotail and in the solar wind. The mission will determine the structure, formation, refilling, and downstream evolution of the lunar wake. The three remaining THEMIS satellites continue to study substorms that are visible in the Northern Hemisphere as aurora borealis.

The Austrian Space Agency (ASA) and the Space Research Institute of the Austrian Academy of Sciences (IWF/ÖAW) developed and tested the fluxgate magnetometer (FGM) electronics, a subsystem of the fluxgate magnetometer instrument, on all five probes. The Canadian Space Agency (CSA), the French National Center for Space Studies (CNES), and the German Aerospace Center (DLR) also contributed to this mission. NASA's contributions include the spacecraft, the launch, the ground network of observatories, and overall mission management.

For more information on THEMIS, please visit *http://www.nasa.gov/mission_pages/themis/main/index.html*.

For more information on ARTEMIS, please visit *http://www.nasa.gov/mission_pages/artemis/index.html*.

Science

An artist's concept of the Juno spacecraft orbiting Jupiter.

Belgium

Juno

NASA's Juno spacecraft, launched in 2011 and slated to arrive in orbit around Jupiter in 2016, will gather data on Jupiter's interior structure, atmospheric composition and dynamics, and polar magnetosphere. This data will be used by scientists to study the origin and evolution of Jupiter, thereby furthering our understanding of planetary and solar-system formation.

Using a spinning, solar-powered spacecraft, Juno will make maps of the gravitational and magnetic fields, auroral emissions, and atmospheric composition of Jupiter from a polar orbit. Juno will carry precise high-sensitivity radiometers, magnetometers, and gravity science systems. Juno's 32 orbits over 11 days will sample Jupiter's full range of latitudes and longitudes. From its polar perspective, Juno will combine in situ and remote sensing observations to explore the polar magnetosphere and determine what drives Jupiter's remarkable auroras. Jupiter's solid core and abundance of heavy metals in the atmosphere make it an ideal model to help improve our understanding of the origin of giant planets. Juno will measure global abundances of oxygen and nitrogen by mapping the gravitational field and using microwave observations of water and ammonia.

The Belgian Federal Science Policy Office (BELSPO) and the Liège Space Center (CSL) have participated in the development of Juno payloads. BELSPO contributed the Scan Mirror Assembly (SMA) to the Ultraviolet Spectrometer (UVS) instrument for the Juno mission. The UVS instrument will remotely sense Jupiter's auroral morphology and brightness, provide a context for the in situ measurements, and map the mean energy and flux of precipitating particles.

For more information on the Juno mission, please visit *http://juno.nasa.gov/* and *http://missionjuno.swri.edu/*.

Belgium

SPP
Solar Probe Plus

NASA's Solar Probe Plus (SPP) mission is a project in the Living With a Star (LWS) program, a series of missions designed to gather critical information about the Sun and its effects on Earth and other planetary systems. By flying into the Sun's outer atmosphere, SPP will gather data on the processes that heat the corona and accelerate the solar wind, solving two fundamental mysteries that have been top-priority science goals for many decades. The Belgian Federal Science Policy Office (BELSPO) and the Liège Space Center (CSL) are collaborating with NASA on the Wide-field Imager for Solar Probe (WISPR) investigation on SPP. France and Germany are also participating in this mission.

NASA plans to launch SPP in 2018 from Cape Canaveral Air Force Station (CCAFS) in Florida. The mission is planned to last almost 7 years, and the spacecraft will eventually come within 3.7 million miles of the Sun's surface, well within the orbit of Mercury and about eight times closer than any spacecraft has ever approached.

The solar panels of the Solar Probe Plus spacecraft are folded into the shadow of its protective shield as the probe gathers data on its approach to the Sun. (Credit: Johns Hopkins University Applied Physics Laboratory)

SPP will study the streams of charged particles that the Sun hurls into space from a vantage point where the processes that heat the corona and produce solar wind actually occur. At its closest approach, SPP will zip past the Sun at 125 miles per second, protected by a carbon-composite heat shield that must withstand up to 2600° Fahrenheit and survive blasts of radiation and energized dust at levels not experienced by any previous spacecraft. The primary scientific objectives to be carried out during the SPP mission include: determining the structure and dynamics of the magnetic fields at the sources of both fast and slow solar wind, tracing the flow of energy that heats the corona and accelerates the solar wind, and determining what mechanisms accelerate and transport energetic particles. Instruments include a wide-field imager, fast-ion analyzer, fast-electron analyzer, energetic-particle instrument, magnetometer, and plasma-wave instrument.

Belgian contributions to the mission include modeling, testing, and evaluating the WISPR investigation on the SPP. The WISPR investigation is a single, visible-light telescope with a wide field of view that will track density fluctuations in the solar corona by imaging visible sunlight scattered by electrons in the corona as the spacecraft traverses through its perihelion passes. Fluctuations can arise from dynamic events, such as coronal mass ejections, and also from the "quiescent" slow and fast solar wind. In addition, the rapid motion of the spacecraft through the corona will result in significant apparent changes due to shifts in its viewpoint, enabling tomographic reconstruction of the structures.

For more information on SPP, please visit *http://solarprobe.gsfc.nasa.gov/* and *http://solarprobe.jhuapl.edu/science/index.php*.

Science

NASA's mobile tracking station in Bermuda provides telemetry, radar, and command and control services.

Bermuda

Mobile Tracking Station

In March 2012, NASA and the government of Bermuda signed an agreement to establish a temporary mobile tracking station in Bermuda. The tracking instrumentation consists of transportable systems provided by NASA's Research Range Services (RRS) program and operated by RRS launch range contractors. NASA anticipates using this temporary mobile tracking station several times per year. NASA's Wallops Flight Facility (WFF) is expected to become the prime mid-Atlantic launch range for a new generation of expendable launch vehicles (ELVs) and resupply missions to the International Space Station (ISS). The Bermuda site is essential for WFF operations to fulfill the considerable set of requirements for current and future NASA missions, including: the use of ELVs to deliver satellites to orbit, the provision of Commercial Resupply Services (CRS) supply missions to the ISS, the launch of extended-duration sounding rockets to orbit for up to 7 days, and the launch of long-range sounding rockets. The Bermuda location provides a 360° unobstructed view that enables tracking of satellites into orbit.

The Bermuda mobile station will acquire vehicle performance data for approximately 5 minutes, starting approximately 4 minutes after lift-off. It will also have range-destruct-command relay capability from WFF. The station consists of modular components such as an electronics control van, several small optical systems, radar and telemetry antennas, and movable power systems that are physically located on Cooper's Island.

In addition to benefiting from both local educational opportunities related to NASA's tracking activities and local contractor work, the government of Bermuda expressed an interest in understanding shore erosion and localized atmospheric changes. As part of the agreement, therefore, WFF will provide Bermuda with information it has developed on the study of shoreline erosion and the use of sensors and data for determining the state of the atmosphere in selected locations with respect to variables such as temperature, moisture, wind velocity, and barometric pressure.

For more information on the mobile tracking station in Bermuda, please visit *http://www.nasa.gov/home/hqnews/2012/mar/HQ_12-074_Bermuda_Agreement.html*.

Brazil

GPM
Global Precipitation Measurement Feasibility Study

NASA and the Brazilian Space Agency (AEB) are conducting a joint scientific and engineering feasibility study of potential cooperation in Global Precipitation Measurement (GPM)–related scientific research, ground validation of GPM satellite data, and other related activities.

The GPM mission is an international network of satellites that provides next-generation global observations of rain and snow. Building upon the success of the Tropical Rainfall Measuring Mission (TRMM), the GPM concept centers on the deployment of a "Core" satellite carrying an advanced radar/radiometer system to measure precipitation from space and serve as a reference standard to unify precipitation measurements from a constellation of research and operational satellites. Through improved measurements of global precipitation, the GPM mission will help to advance our understanding of Earth's water and energy cycle, improve forecasting of extreme events that cause natural hazards and disasters, and extend current capabilities in using accurate and timely information of precipitation to directly benefit society.

The GPM mission is a multisatellite constellation project that is being jointly developed by NASA and the Japan Aerospace Exploration Agency (JAXA). The data acquired by the GPM mission will be beneficial for monitoring and predicting climatological and meteorological changes and for improving the accuracy of weather and precipitation forecasts. The GPM Core Observatory launched February 2014.

For more information on the GPM mission, please visit *http://pmm.nasa.gov/precipitation-measurement-missions*.

An artist's concept of the GPM core observatory and other satellites in the GPM network orbiting Earth.

Science

Brazil

Ozone Cooperation

NASA and the Brazilian Space Agency (AEB) are working together to study the concentrations of various atmospheric constituents in order to contribute to an improved understanding of the Earth's ozone layer. This project supplements measurements being made from NASA's Wallops Flight Facility (WFF), Virginia, and other sites for coverage of high Earth latitudes.

Measurements made from Natal, Brazil, provide near-equatorial data needed for coverage of low Earth latitudes. Natal is one of 14 observation sites for collection of these data. The precise location of the site is 40 kilometers north of the city of Natal in the small seaside village of Maxaranguape. Brazil has utilized balloon-borne ozonesonde assets in Natal since 1978.

These measurements are supplemented by ozonesonde data obtained from other NASA-AEB cooperative projects, including the NASA Southern Hemisphere ADditional OZonsondes (SHADOZ) project. The purpose of SHADOZ is to profile the climatology of tropical ozone in the equatorial zone and to validate and improve satellite remote sensing techniques for estimating tropical ozone. AEB's responsibilities for this activity are carried out through the Brazilian National Institute for Space Research (INPE).

Fore more information on NASA's ozone study activities, please visit *http://croc.gsfc.nasa.gov/shadoz/*.

For access to ozone study data from Wallops Flight Facility and INPE, please visit *http://uairp.wff.nasa.gov/*.

A balloon is prepared for a scheduled flight from a launch site at a seaside village called Maxaranguape, located 40 kilometers north of the city of Natal, Brazil.

Canada

Aircraft Icing Research

The aviation industry has experienced numerous engine power-loss events associated with ice crystal formation in high-altitude convective weather, often in a warm tropical environment. As a result, the aviation industry has identified the need for detailed measurements of the fundamental behavior of ice crystal formation on modern gas turbine engines in warm-air environments. NASA is partnering with the Canadian National Research Council (NRC) to conduct rig-test experiments in ground-based facilities to improve understanding of ice crystal accretion physics on heated surfaces, which are representative of gas turbine components.

NASA's collaboration with the NRC will support the development of ground-based simulation test capabilities for both organizations. These simulation capabilities will in turn be used to shape future regulatory requirements for aircraft that fly in environments that are likely to produce ice crystals in aircraft engines. The research will be conducted over a multiyear period at facilities located in both the United States and Canada.

NASA and NRC will also cooperate in the multiyear High Ice Water Content (HIWC) field campaign and conduct joint research to improve the measurement capabilities of high ice-water content and mixed phase (liquid and ice) environments. Cooperation in HIWC and joint icing research will support modifications to existing atmospheric instruments and the development of isokinetic probes used for ground-based and flight-based measurements.

For more information on NASA's icing research, please visit *http://www.nasa.gov/topics/aeronautics/features/preps_solve_engine_icing.html*.

Technicians install a series of horizontal spray bars inside an opening to a NASA test facility that will simulate engine-icing conditions on the ground.

More than 20 meteorological instruments outside and inside the Gulfstream aircraft will be used to collect data on engine-icing conditions.

Science

Canada

CloudSat

An artist's concept of NASA's CloudSat spacecraft, which will provide the first global survey of cloud properties to better understand their effects on weather and climate.

NASA and the Canadian Space Agency (CSA) jointly developed the Cloud Profiling Radar for NASA's CloudSat satellite that was launched in April 2006. CloudSat measures the vertical structure of clouds from space. It is the first satellite to fly millimeter-wave radar that is capable of seeing a large fraction of clouds and precipitation, ranging from very thin cirrus clouds to thunderstorms producing heavy precipitation. CloudSat collects data that is needed to evaluate and improve the way in which clouds are represented in global climate models, thereby contributing to better predictions of cloud formations and a more detailed understanding of their role in both climate change and cloud-climate feedback.

CloudSat was launched together with the French National Center for Space Studies (CNES) Cloud-Aerosol LIDAR (light detection and ranging) and Infrared Pathfinder Satellite Observations (CALIPSO) spacecraft. It is flying in orbital formation as part of the Afternoon Constellation (A-Train) of Earth science satellites, including NASA's Aqua and Aura and CNES's Polarization and Anisotropy of Réflectances for Atmospheric Sciences coupled with Observations from a LIDAR (PARASOL) mission. CloudSat maintains a tight formation with CALIPSO in order to overlap measurement footprints at least 50 percent of the time.

For more information on CloudSat, please visit *http://cloudsat.atmos.colostate.edu/*.

Canada

ISS
International Space Station

The Canadian Space Agency's (CSA) primary contribution to the International Space Station (ISS) Program is the Mobile Servicing System (MSS), which consists of the Mobile Base System, Space Station Remote Manipulator System (SSRMS), and the Special Purpose Dexterous Manipulator (Dextre).

The ISS Program is a partnership among the space agencies of Canada, Europe, Japan, Russia, and the United States. International crews have continuously inhabited the ISS since November 2000. Cargo transportation is provided by European, Russian, and Japanese vehicles and by NASA's contracts for Commercial Resupply Services (CRS). Crew transportation and rescue is provided by the Russian Federal Space Agency and, beginning in 2017, by NASA's commercial crew transportation services.

Astronaut Rick Linnehan conducts a spacewalk during the STS-123 mission while anchored to a Canadarm2 mobile foot restraint. During the spacewalk, Linnehan helped install the tool-change-out mechanisms on the Canadian-built Dextre robotic system, the final element of the ISS Mobile Servicing System.

NASA's contributions to the ISS Program include overall management, coordination, and integration of the ISS Program across the partnership for operations, safety, transportation, and integrated hardware performance. NASA hardware components include the U.S. laboratory module, the airlock for extravehicular activities (EVA), truss segments to support the U.S. solar arrays and thermal radiators, three connecting nodes, and living quarters.

The ISS functions as an orbital microgravity and life sciences laboratory, a test bed for new technologies in areas such as life support and robotics, and a platform for astronomical and Earth observations. The ISS also serves as a unique engineering test bed for flight systems and operations critical to NASA's future exploration missions. U.S. research on the ISS concentrates on the long-term effects of space travel on humans and on technology development activities in support of exploration. The ISS is a stepping-stone for human exploration and scientific discovery beyond low-Earth orbit.

In the grasp of the ISS robotic Canadarm2, the H-II Transfer Vehicle 3 (HTV-3) Exposed Pallet is moved for installation on the Japan Aerospace Exploration Agency (JAXA) HTV-3 while it is docked to the Space Station in July 2012.

CSA's MSS played a key role in the on-orbit construction of the ISS and continues to play a key role in general ISS operations and maintenance. The MSS is comprised of three main elements: the SSRMS, called Canadarm2—a 17-meter-long robotic arm with seven motorized joints; the Mobile Base System—a moveable work platform that transports the Canadarm2 along rails covering the length of the ISS, serves as a storage facility for astronauts during space walks, and can serve as a base for both the Canadarm2 and Dextre simultaneously; and Dextre—a smaller, 3.5 meter-long, two-arm robot that can perform delicate maintenance and servicing tasks on the exterior of the ISS, where precise handling is required. CSA's MSS Operation Complex, located at CSA Headquarters in St. Hubert, Quebec, supports the operation and maintenance of the MSS and is the main facility for MSS mission planning, equipment monitoring, and training.

For more information on CSA's contributions to the ISS, please visit *http://www.asc-csa.gc.ca/eng/iss/default.asp* and *http://www.nasa.gov/mission_pages/station/main/index.html*.

Science

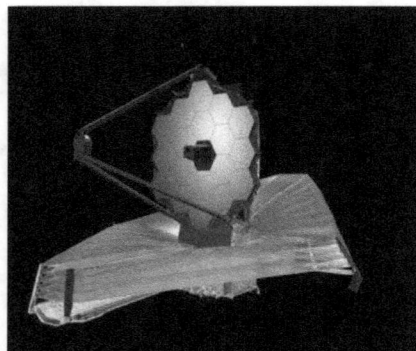

An artist's concept of the JWST.

Designed to block solar light and keep the JWST observatory operating at cryogenic temperature, the five-layer sunshield consists of thin membranes made from a polymer-based film and supporting equipment such as spreader bars, booms, cabling, and containment shells.

Canada

JWST
The James Webb Space Telescope

The James Webb Space Telescope (JWST), planned for launch in 2018, will be a large infrared telescope with a 6.5-meter primary mirror. JWST will enable the study of every phase in the history of our universe, ranging from the first luminous glows after the Big Bang, to the formation of solar systems capable of supporting life on Earth-like planets, to the evolution of our own solar system. JWST is an international collaboration between NASA, the Canadian Space Agency (CSA), and the European Space Agency (ESA). CSA is providing the Fine Guidance Sensor (FGS)/Near-Infrared Imager and Slitless Spectrograph (NIRISS) for the telescope.

Referred to as the JWST Observatory, the space-based portion of the JWST system consists of three parts: the Optical Telescope Element (OTE), the Spacecraft Element, and the Integrated Science Instrument Module (ISIM). The OTE is the eye of the JWST, collecting light from space and providing it to scientific instruments. The Spacecraft Element is composed of the spacecraft bus (which provides support functions for the observatory) and the sunshield system (which separates the observatory into a warm Sun-facing side and a cold side that faces away from the Sun). The ISIM contains four instruments: the Mid-Infrared Instrument (MIRI), the Near-Infrared Spectrograph (NIRSpec), the Near-Infrared Camera (NIRCam), and the FGS/NIRISS. JWST's instruments will be designed to work primarily in the infrared range of the electromagnetic spectrum, with some capability in the visible range.

Several innovative technologies have been developed for JWST, including a folding, segmented primary mirror that is adjusted to shape after launch; ultra-lightweight beryllium optics; detectors capable of recording extremely weak signals; microshutters that enable programmable object selection for the spectrograph; and a cryocooler for cooling the mid-infrared detectors to the required 7° kelvin operating temperature.

The CSA-provided FGS is a broadband guider camera that is used for both "guide star" acquisition and fine pointing. Its field of view is sufficient to provide a 95-percent probability of acquiring a guide star for any valid pointing direction. The FGS will provide mission-critical support for the JWST Observatory's attitude control system. NASA is responsible for the design and development of the JWST Observatory and for overall management of the mission.

For more information on JWST, please visit *http://jwst.gsfc.nasa.gov/*.

Canada

MSL
Mars Science Laboratory

NASA's Mars Science Laboratory (MSL) mission utilized an innovative sky crane to set down the Curiosity rover onto the surface of Mars in August 2012, where it began its mission to assess whether Mars has ever had an environment capable of supporting small life forms called microbes. The MSL mission, launched by NASA in November 2011, is planned to last at least one Martian year (687 Earth days), but it has the potential to operate for much longer.

This artist's concept features NASA's MSL Curiosity rover, a mobile robot for investigating Mars's past and/or present ability to sustain microbial life. (Credit: NASA/Jet Propulsion Laboratory–Caltech)

The primary scientific objective, to be carried out during the surface science phase of the MSL mission, is to assess the biological potential of at least one target area by characterizing the local geology and geochemistry, investigating planetary processes relevant to habitability (including the role of water), and characterizing the broad spectrum of surface radiation. The landing site, Gale Crater, was selected based on an assessment of safety and planetary protection and an analysis by the scientific community. Since its landing, MSL has already begun to make new and exciting science discoveries, such as observing a natural intersection of three kinds of terrain at an area called Glenelg and finding evidence that a stream once ran across the area on Mars where the rover is driving. Although earlier missions detected the presence of water on Mars, the images taken by MSL of rocks that contain ancient streambed gravel are the first of their kind.

The Canadian Space Agency (CSA) provided MSL's Alpha Particle X-ray Spectrometer (APXS), which will measure the abundance of chemical elements in rocks and soils. The instrument will be placed in contact with rock and soil samples on Mars in order to expose the material to alpha particles and x rays that are emitted during the radioactive decay of the element curium. Scientists will use the APXS to help characterize and select rock and soil samples by examining the interiors of the rocks revealed through abrasion. By analyzing the elemental composition of rocks and soils, scientists hope to understand how the material first formed and whether it was later altered by wind, water, or ice. During the first use of the robotic arm to "touch" a rock, the MSL team used the APXS to study the rock—called Jake Matijevic—and to identify its chemical make-up. Earlier versions of this APXS flew on NASA's Mars Pathfinder and Mars Exploration Rover missions.

For more information on the MSL mission and the APXS instrument, please visit *http://mars.jpl.nasa.gov/msl/* and *http://marsprogram.jpl.nasa.gov/msl/mission/instruments/spectrometers/apxs/*.

Science

An artist's concept of OSIRIS-REx approaching the Bennu asteroid. The mission seeks to return a sample that will help scientists investigate planet formation and the origin of life; the data collected at the asteroid will also aid our understanding of asteroids that can impact Earth. (Credit: NASA/Goddard Space Flight Center/University of Arizona)

Canada

OSIRIS-REx
Origins Spectral Interpretation Resource Identification Security–Regolith Explorer

The Origins Spectral Interpretation Resource Identification Security–Regolith Explorer (OSIRIS-REx), planned for launch in 2016, is the first U.S. mission developed to return a sample from an asteroid to Earth. The OSIRIS-REx mission is designed to return a minimum of 60 grams of bulk regolith and 26 square centimeters of surface material from the asteroid. Analysis of these samples could provide unprecedented insight into pre-solar history through the initial stages of planet formation and up to the origin of life. NASA is collaborating with the Canadian Space Agency (CSA) and the French National Center for Space Studies (CNES) on this mission.

The mission's target asteroid is an accessible volatile- and organic-rich remnant from the early solar system. In April 2013, the asteroid was named "Bennu" as a result of an international naming contest. Bennu is the name of an Egyptian deity that is usually depicted as a heron. This name was chosen for the asteroid because OSIRIS-REx's Touch-and-Go Sample Mechanism (TAGSAM) arm and solar panels resemble the neck and wings of Bennu.

Prior to sample acquisition, OSIRIS-REx will perform comprehensive global mapping of the texture, mineralogy, and chemistry of Bennu, thereby resolving geological features, revealing its geologic and dynamic history, and providing context for the returned samples. Mission scientists will also study the Yarkovsky effect, whereby sunlight striking a rotating body in space can, over time, exert enough force on the object to change its orbit.

OSIRIS-REx is scheduled to rendezvous with Bennu in 2019, and the sample return capsule should land on Earth in 2023. The OSIRIS-REx science instrument suite is expected to consist of a camera, a laser altimeter, a visible and infrared spectrometer, a thermal emission spectrometer, and a regolith x-ray imaging spectrometer.

CSA is providing the OSIRIS-REx Laser Altimeter (OLA), which has two separate transmitter assemblies to provide high-resolution topographical information. OLA, whose science team is led by the University of Calgary, will scan the entire surface of the asteroid to create a highly accurate, three-dimensional model that will provide mission scientists with fundamental and unprecedented information on the asteroid's shape, topography, surface processes, and evolution. This instrument will also create local maps to assist scientists in selecting the best sites for sample collection.

For more information on OSIRIS-Rex, please visit *http://science.nasa.gov/missions/osiris-rex/*, *http://www.atmosp.physics.utoronto.ca/MOPITT/home.html*, and *http://osiris-rex.lpl.arizona.edu/*.

Canada

RPM
Resource Prospector Mission

NASA is cooperating with the Canadian Space Agency (CSA) to develop a mission to characterize the presence of water and other volatiles on the lunar surface by direct ground-truth measurements. The mission will also validate the feasibility of extracting volatiles and oxygen from the lunar regolith and serve as a precursor for future prospecting missions to a variety of locations around the solar system.

NASA has been working on a Resource Prospector Mission (RPM) payload, an advanced instrument package designed to be mounted on a rover and driven over the surface of the Moon to map the distribution of water ice and other useful compounds.

In July 2012, on the Mauna Kea volcano in Hawaii, NASA, with the cooperation of CSA, demonstrated the ability of the RPM payload mounted atop a Canadian rover to be maneuvered in a lunar-like landscape and perform the tasks necessary to map the distribution of water ice and other useful compounds on the lunar surface. With CSA's coring drill, avionics, and rover, the Hawaii field demonstration prepared RPM for a possible future mission to the Moon.

Current collaborations with CSA include engineering studies for a coring drill for the RPM mission and analyses on rover operations concepts. CSA has contemplated contributing a rover to the flight mission, and studies are proceeding. Launch of an RPM mission could occur as early as 2018.

For more information on the 2012 Field Demonstration, please visit *http://www.nasa.gov/exploration/systems/ground/resolverover.html*.

The Resource Prospector Mission is a miniature drilling and chemistry plant packaged onto a medium-sized rover to collect and analyze volatiles in lunar regolith.

Canada

RRM
Robotic Refueling Mission

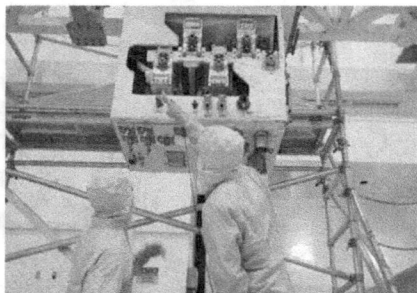

The RRM module is prepared for launch on STS-135 at NASA's Kennedy Space Center.

Spacewalker Mike Fossum, while anchored to the Canadian Space Agency's Dextre robot, carries the RRM module from the Atlantis cargo bay to a temporary platform on the International Space Station.

The Robotic Refueling Mission (RRM) is a joint NASA–Canadian Space Agency (CSA) technology demonstration that CSA conducted on the International Space Station (ISS). The RRM payload is located on an external pallet on the ISS. It consists of task boards that represent the external interfaces of a legacy satellite and specialized tools that attach and are used by CSA's Special Purpose Dexterous Manipulator (Dextre).

Dextre utilizes the four unique RRM tools to demonstrate a suite of satellite-servicing and refueling tasks, including cutting and manipulating protective blankets and safety wires, unscrewing caps and accessing valves, and transferring fluid. RRM performs tasks to demonstrate the ability to robotically refuel and repair satellites in orbit that were not designed to be serviced. This is the first use of Dextre for technology research and development; previously it had been used only for robotic maintenance of the ISS.

The RRM payload was launched on Space Transportation System (STS)–135 in July 2011 and began operations in March 2012. RRM operations are entirely remote-controlled by flight controllers at NASA's Goddard Space Flight Center (GSFC) in Greenbelt, Maryland; Johnson Space Center (JSC) in Houston, Texas; Marshall Space Flight Center (MSFC) in Huntsville, Alabama; and CSA's control center in St. Hubert, Quebec.

Robotic refueling and servicing could extend a satellite's lifespan, potentially offering satellite owners and operators years of additional service and revenue, more value from the initial satellite investment, and significant savings in delayed replacement costs.

For more information on this joint NASA-CSA mission, please visit *http://www.nasa.gov/mission_pages/station/research/experiments/RRM.html* and *http://www.asc-csa.gc.ca/eng/iss/rrm/default.asp*.

Canada

SCISAT-1

NASA and the Canadian Space Agency (CSA) continue to cooperate on CSA's SCISAT-1 mission, which was launched in 2003 from Vandenberg Air Force Base (VAFB), California. In addition to providing the launch on a Pegasus rocket, NASA contributed to the development of the Atmospheric Chemistry Experiment (ACE)–Fourier Transform Spectrometer (FTS) instrument. NASA also provided science team support and algorithm development for analyzing the data.

The overall objective of the SCISAT-1 mission is to improve understanding of the depletion of the ozone layer, with close attention paid to the layer over Canada and the Arctic. Scientists combined the measurements obtained by the ACE-FTS and the Canadian Measurements of Aerosol Extinction in the Stratosphere and Troposphere Retrieved by Occultation (MAESTRO) instruments (also flying on SCISAT-1) with data gathered by ground-based, balloon-based, and other space-based missions in order to predict future trends relating to the chemistry of the ozone layer and its dynamics. SCISAT-1 provides near-global coverage, with observations of gas constituents, aerosols, and clouds in the stratosphere, as well as observations in the troposphere. SCISAT-1's high accuracy and fine vertical resolution as it probes the Earth's atmosphere enables it to validate measurements from NASA's Aura mission, which also measures atmospheric composition.

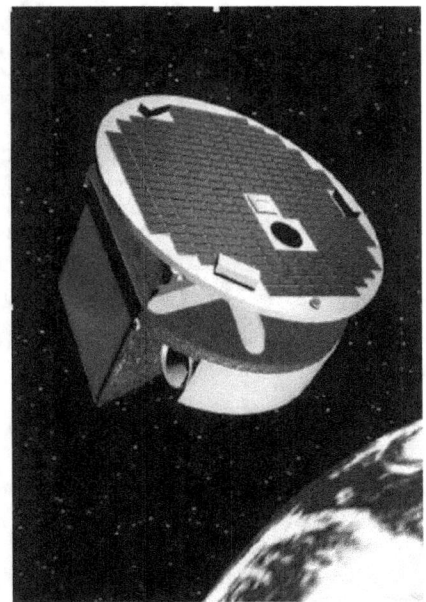

An artist's concept of the SCISAT-1 spacecraft.

In conjunction with other NASA and international partner instruments and missions that probe the Earth's atmosphere, SCISAT-1 provides important data to better understand the chemistry and dynamics of the atmosphere that affect Earth's protective ozone layer and their relationship to climate change.

For more information on SCISAT-1, please visit *http://www.asc-csa.gc.ca/eng/satellites/scisat/default.asp*.

Science

Canada

SMAP
Soil Moisture Active Passive

An artist's concept of the SMAP satellite in orbit; the SMAP instrument incorporates an L-band radar and an L-band radiometer, which enable night or day observations of soil moisture through moderate vegetation cover, independent of cloud cover.

NASA's Soil Moisture Active Passive (SMAP) mission is one of the four tier-1 Earth science missions recommended by the U.S. National Research Council's Earth Science Decadal Survey. SMAP will give scientists a more detailed look at the intersection of the water, energy, and carbon cycles on the land surface, enhancing our understanding of the climate system.

Information from SMAP will be used to improve weather forecasting and short-term climate predictions, while also aiding flash flood guidance and drought assessment. The SMAP instrument suite will consist of a radiometer and a nonimaging synthetic aperture radar (SAR). SMAP measurements are expected to yield estimates of soil moisture and freeze/thaw state. The SMAP mission is designed to acquire these measurements for a period of 3 years and to use a comprehensive validation program to assess random errors and regional biases in the soil moisture and freeze/thaw estimates.

SMAP will employ a dedicated spacecraft with an instrument suite that is planned for launch into a near-polar, Sun-synchronous orbit no earlier than 2014.

The Canadian Space Agency (CSA) is supporting a Canadian science team in pre- and post-launch data calibration, validation, and algorithm development activities.

For more information on SMAP, please visit *http://science.nasa.gov/missions/smap/* and *http://smap.jpl.nasa.gov/*.

Canada

Terra

The Canadian Space Agency (CSA) provided the Measurements of Pollution in the Troposphere (MOPITT) instrument on NASA's Terra satellite, which was launched in 1999. MOPITT measures the pollution of Earth's atmosphere from space using gas correlation spectroscopy. Terra, which had a planned mission life of 6 years, is operating in extended mission phase.

MOPITT is Canada's first major instrument to measure the pollution of Earth's atmosphere from space, as well as the first satellite sensor to use gas correlation spectroscopy. The sensor measures emitted and reflected radiance from Earth in three spectral bands to determine the concentration of carbon monoxide and methane. MOPITT's spatial resolution is 22 kilometers at nadir, and it views Earth in swaths that are 640 kilometers wide.

MOPITT's continuous scan of the atmosphere has provided the world with the first long-term global measurements of carbon monoxide and methane gas levels in the atmosphere. These data, together with the other sensor measurements, are helping to form continuing integrated measurements of land, air, water, and life processes. These measurements are used by scientists to predict the long-term effects of pollution, to understand the increase of ozone in the lower atmosphere, and to guide the evaluation and application of shorter-term pollution controls.

For more information on MOPITT, please visit *http://terra.nasa.gov/*, *http://www.atmosp.physics.utoronto.ca/MOPITT/home.html*, and *http://www.eos.ucar.edu/mopitt/*.

An illustration of how the MOPITT instrument scans a swath of Earth's atmosphere.

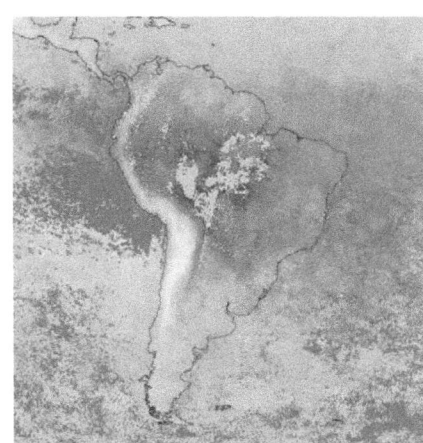

This image of South America shows satellite observations of carbon monoxide, measured by the MOPITT sensor on the Terra satellite, as intense fires burned across the continent in September 2007. Carbon monoxide is one of several gases released by burning vegetation. (Credit: NASA image created by Jesse Allen, Earth Observatory, using data provided by the National Center for Atmospheric Research and the University of Toronto)

Science

Canada

THEMIS
Time History of Events and Macroscale Interactions during Substorms
and
ARTEMIS
Acceleration, Reconnection, Turbulence and Electrodynamics of the Moon's Interaction with the Sun

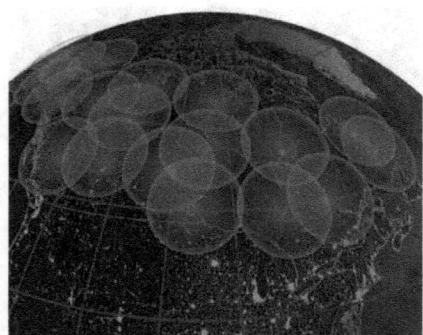

In addition to the 5 THEMIS spacecraft launched into space, 20 THEMIS ground stations located across North America can observe aurora from the ground.

NASA's Time History of Events and Macroscale Interactions during Substorms (THEMIS) mission originated as a 2-year mission consisting of five identical satellites, or probes, launched together in 2007 to study the violent and colorful eruptions in auroras. The mission incorporates a network of ground-based auroral observatories.

THEMIS helps to determine which physical processes in near-Earth space initiate violent substorm eruptions in the Earth's magnetosphere. Substorms intensify auroras and create a dramatic "dancing" effect in them. Aligning five identical probes over observatories on the North American continent has allowed scientists to collect coordinated measurements along the Earth's magnetic field lines, thereby providing the first comprehensive look at the onset of substorms and the manner in which they trigger auroral eruptions.

In 2009, NASA assigned two of the THEMIS satellites to a new mission—the Acceleration, Reconnection, Turbulence and Electrodynamics of the Moon's Interaction with the Sun (ARTEMIS) mission—to measure solar wind turbulence at the Moon. Having repositioned two of the five THEMIS probes in coordinated, lunar-equatorial orbits, ARTEMIS is now performing the first systematic, two-point observations of the distant magnetotail, the solar wind, and the lunar space environment. The primary objectives of the mission are to study how particles are accelerated at reconnection sites and shocks and how turbulence develops and evolves in Earth's magnetotail and in the solar wind. The mission will determine the structure, formation, refilling, and downstream evolution of the lunar wake. The three remaining THEMIS satellites continue to study substorms that are visible in the Northern Hemisphere as aurora borealis.

The German Aerospace Center (DLR) and the Technical University of Braunschweig developed and tested the fluxgate magnetometer (FGM) installed on each probe. The Austrian Space Agency (ASA), the Space Research Institute of the Austrian Academy of Sciences (IWF/ÖAW), the Canadian Space Agency (CSA), and the French National Center for Space Studies (CNES) also contributed to this mission. NASA's contributions include the spacecraft, the launch, the ground network of observatories, and overall mission management.

For more information on THEMIS, please visit *http://www.nasa.gov/mission_pages/themis/main/index.html*.

For more information on ARTEMIS, please visit *http://www.nasa.gov/mission_pages/artemis/index.html*.

Czech Republic

Van Allen Probes

The Van Allen Probes mission, part of NASA's Living With a Star (LWS) program, is providing unprecedented insight into the physical dynamics of Earth's radiation belts and giving scientists the data they need to make predictions of changes in this critical region of space. The mission features two spacecraft, launched in 2012, that orbit Earth and sample the harsh radiation belt environment where major space weather activity occurs and many spacecraft operate. These two identical spacecraft are measuring the particles, magnetic and electric fields, and waves that fill geospace. By having two spacecraft take identical measurements while following the same path, scientists can begin to understand how the belts change in both space and time.

An artist's concept of the twin Van Allen Probes spacecraft.

The Van Allen Probes are designed to help understand the Sun's influence on Earth and in near-Earth space by studying the planet's radiation belts on various scales of space and time. The Van Allen Probes spacecraft will allow scientists to study the Van Allen radiation belts that surround our planet and to learn more about the processes that create them and cause them to vary in size and intensity. The goal of the mission is to understand the acceleration, global distribution, and variability of energetic electrons and ions in the radiation belts. Understanding the radiation belt environment and its variability has important practical applications in the areas of spacecraft operations, spacecraft and spacecraft system design, mission planning, and astronaut safety.

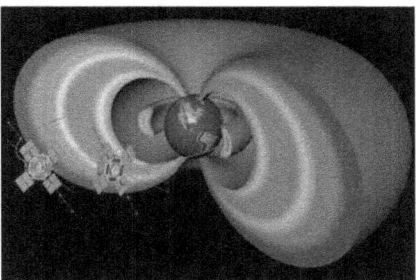

An artist's rendering showing the two Van Allen Probes observing the Sun and its effects on Earth. (Credit: Johns Hopkins University Applied Physics Laboratory)

The instruments on the two Van Allen Probes spacecraft will provide the measurements needed to characterize and quantify the processes that produce relativistic ions and electrons. They will measure the properties of the charged particles that comprise the Earth's radiation belts and the plasma waves that interact with them, the large-scale electric fields that transport them, and the magnetic field that guides them.

Charles University in Prague is collaborating on the flight software for the waveform receivers (known as "Waves"), which are part of the Electric and Magnetic Field Instrument Suite and Integrated Science (EMFISIS) instrument suite. The EMFISIS instrument suite will help scientists to better understand the origin and role of plasma waves in particle acceleration and in the evolution of the magnetic field.

The Institute of Atmospheric Physics (IAP), as part of the Academy of Sciences of the Czech Republic, is cooperating with NASA to share heliophysics and space weather research data from new NASA missions such as the Van Allen Probes. The IAP will provide the necessary ground assets to acquire and process the space weather broadcast data from the Van Allen Probes and will share this dataset with NASA researchers.

For more information on the Van Allen Probes, please visit *http://science.nasa.gov/missions/van-allen-probes/*.

Education and Outreach

Japan Aerospace Exploration Agency (JAXA) astronaut Satoshi Furukawa presents his completed LEGO model of the International Space Station.

NASA and LEGO sponsor a "Build the Future" activity at NASA's Kennedy Space Center in November 2010.

Denmark

LEGO

NASA's cooperation with the LEGO Group formally began with the signing of an international Space Act Agreement in October 2010. This relationship has served to develop innovative public education products and outreach activities, further NASA's missions in outreach and education, and highlight the shared theme of "Building and Exploring Our Future."

Since the start of the partnership between LEGO and NASA, many projects have been undertaken to encourage interest and awareness in NASA's science, technology, engineering, and exploration programs. In fall 2010, during the pre-launch activities of Space Transportation System (STS)–133, an activity was set up for children to build their vision of the future in space using over 100,000 LEGO bricks at NASA's Kennedy Space Center (KSC). This "Build the Future" event was repeated at the subsequent NASA Space Shuttle and rocket launches at KSC of STS-134, STS-135, Juno, and the Mars Science Laboratory.

To help promote NASA's educational goals related to science, technology, engineering, and mathematics (STEM) education, flight-certified LEGO bricks were sent to the International Space Station (ISS) in February 2011 aboard STS-134. The astronauts used the educational kits to demonstrate to children the differences in building things in the microgravity environment of space compared to on Earth.

LEGO created toy figures (LEGO Minifigures) of Galileo and the Roman gods Jupiter and Juno that were affixed to the outside of NASA's Juno spacecraft prior to its launch in August 2011. The LEGO Minifigures were developed to help spark public interest in the mission and in the study of the planet Jupiter. When the Juno spacecraft arrives at its destination, the LEGO Minifigures will be the furthest toys from Earth.

LEGO Mindstorms kits, which include the components to build and program robots made from LEGO bricks, are also planned to help further involve the public in NASA missions. The kits are intended to help middle-school teachers learn how to teach robotics in the classroom, but they can also be used by extracurricular groups. The kits will include components related to a manned trip to Mars and other locations, with inductive research ideas based on NASA's three Space Technology Grand Challenges.

For more information on NASA's cooperation with LEGO, please visit *http://www.nasa.gov/audience/foreducators/nasa-lego-partnership.html*, *http://www.nasa.gov/mission_pages/shuttle/behindscenes/lego.html*, and *http://www.legospace.com/en-us/Default.aspx.*

Denmark

MER
Mars Exploration Rovers

The Niels Bohr Institute for Astronomy, Physics, and Geophysics (NBI) at the University of Copenhagen provided the permanent magnet arrays for the Athena Payload on NASA's Mars Exploration Rover (MER) mission. Athena is MER's science payload, consisting of a package of science instruments that includes two instruments for surveying the landing site and three instruments located on each rover's arm for close-up study of rocks. Each rover has three sets of permanent magnet arrays that work in conjunction with the science instruments and collect airborne dust for analysis. A periodic examination of this dust can reveal clues about its mineralogy and the planet's geological history.

An artist's concept of an MER on the Martian surface.

The MER mission, involving two robotic geologists—Spirit and Opportunity—is part of NASA's Mars Exploration Program. Primary among the MER mission's scientific goals is to search for and characterize a wide range of rocks and soils that hold clues to past water activity. The MER mission seeks to determine where conditions may once have been favorable to life by analyzing the climate and water histories at sites on Mars. Each rover is equipped with the Athena science payload, which is used to read the geological record at each site, investigate what role water played there, and determine how suitable the conditions would have been for life.

Spirit and Opportunity worked on Mars well beyond their originally planned 3-month missions; Spirit fell silent in March 2010, but Opportunity has continued to send data downlinks to Earth. Within 2 months of landing on Mars, Opportunity found geological evidence of an environment that was once wet.

This image was captured by the Spirit Rover on January 15, 2004, as it rolled off its lander platform onto the Martian soil.

In 2011, Opportunity discovered bright veins of a mineral deposited by water on the Martian surface. This water-related mineral, believed to be gypsum, supports the hypothesis that water flowed through underground fractures in the rock. The vein is made up of calcium sulfate that was identified by Opportunity's Alpha Particle X-ray Spectrometer (APXS).

NASA's Opportunity rover is one of five active robotic missions at Mars, including NASA's Mars Odyssey orbiter and Mars Reconnaissance Orbiter (MRO), the European Space Agency's (ESA) Mars Express orbiter, and NASA's Mars Science Laboratory (MSL). The orbiters and surface missions complement each other in many ways. Observations by the rovers provide ground-level understanding for interpreting global observations made by the orbiters. In addition to their own science missions, the orbiters relay data from the Mars rovers.

For more information on MER, please visit *http://marsrovers.nasa.gov/home/index.html* and *http://athena.cornell.edu/*.

Science

Denmark

NuSTAR
Nuclear Spectroscopic Telescope Array

NASA's Nuclear Spectroscopic Telescope Array (NuSTAR) mission, launched in June 2012, will allow astronomers to study the universe in high-energy x rays. Nine days after its launch, engineers at NuSTAR's mission control at University of California Berkeley sent a signal to the spacecraft to begin deploying its 10-meter mast. The mast was deployed and secured over a span of 26 minutes, making it the first deployed mast ever used on a space telescope. As a result of this achievement, NuSTAR became the first focusing hard x-ray telescope to orbit Earth, and it is expected to greatly exceed the performance of the largest ground-based observatories that have examined this region in the electromagnetic spectrum. NuSTAR will also complement astrophysics missions that explore the cosmos in other regions of the spectrum. NASA is collaborating with the Danish Technical University and Italian Space Agency (ASI) on the NuSTAR mission.

NuSTAR is able to focus the high-energy x-ray light into sharp images because of a complex, innovative telescope design. High-energy light is difficult to focus because it reflects off mirrors only when hitting at nearly parallel angles. NuSTAR ingeniously solves this problem by using nested shells of mirrors in its design. It has the most nested shells ever used in a space telescope—133 in each of two optic units. The mirrors were molded from ultrathin glass similar to that found in laptop screens and glazed with even thinner layers of reflective coating.

NuSTAR's optics were constructed by a team that included the Danish Technical University, NASA Goddard Space Flight Center (GSFC), NASA Jet Propulsion Laboratory (JPL), and the Lawrence Livermore National Laboratory (LLNL). The reflective coating was applied in Copenhagen, Denmark. ASI is providing the mission with the use of the Malindi Ground Station, operated by ASI and located in Malindi, Kenya, as the primary NuSTAR data downlink and command uplink facility.

For more information on NuSTAR, please visit *http://www.nasa.gov/nustar*.

An artist's concept of NuSTAR orbiting Earth; NuSTAR has a 10-meter mast that deploys after launch to separate the optics modules (right) from the detectors in the focal plane (left). The spacecraft and the solar panels are with the focal plane. NuSTAR has two identical optics modules in order to increase sensitivity. The background is an image of the Galactic Center obtained with the Chandra X-ray Observatory. (Credit: NASA/Jet Propulsion Laboratory–Caltech)

High-energy x-ray data from NASA's NuSTAR spacecraft has been translated to the color magenta and superimposed on a visible-light view of the Caldwell 5 spiral galaxy. NuSTAR is the first orbiting telescope to take focused pictures of the cosmos in high-energy x-ray light. The visible-light image is from the Digitized Sky Survey. (Credit: NASA/Jet Propulsion Laboratory–Caltech/Digitized Sky Survey)

Europe

Cassini-Huygens

Cassini-Huygens—a joint mission by NASA, the European Space Agency (ESA), and the Italian Space Agency (ASI) to explore Saturn, Titan, and the other moons of the Saturnian system—has two distinct elements: NASA's Cassini orbiter and ESA's Huygens probe. ASI provided the high-gain antenna (HGA) and instrumentation for the Cassini orbiter. In addition, scientists from the United States and Europe contributed to and participated in the science teams that designed and developed several of the scientific instruments for Cassini and the Huygens probe.

In 2005, the Huygens probe landed on Titan, Saturn's enigmatic moon, and sent back the first-ever images from beneath Titan's thick cloud layers. This artist's rendering is based on those images. The parachute that slowed Huygens's reentry is seen in the background, still attached to the lander. (Credit: European Space Agency)

The Cassini-Huygens spacecraft was launched in 1997, and 7 years later it entered the Saturnian system and established orbit around Saturn. In December 2004, the Huygens probe was jettisoned from the Cassini orbiter and successfully completed its 20-day mission to enter the atmosphere and land on the surface of Titan.

A total of seventeen nations contributed to building Cassini-Huygens, and nearly 300 scientists from the United States and Europe study the several gigabytes of data generated by the spacecraft on a daily basis. The following European countries cooperated with NASA on Cassini's scientific instruments and on two of the Huygens probe's scientific instruments, either through provision of hardware or sponsorship of scientists on the instrument science teams that were responsible for the design, development, testing, integration, and post-launch operation of the instruments: Austria, Finland, France, Germany, Hungary, Norway, Sweden, and the United Kingdom.

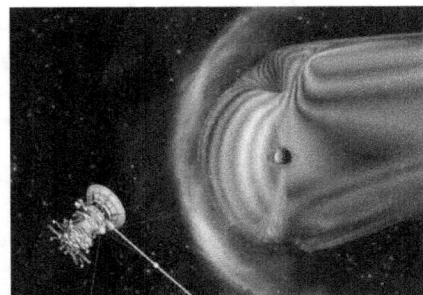

This artist's concept shows NASA's Cassini spacecraft exploring the magnetic environment of Saturn. Saturn's magnetosphere is depicted in gray, while the complex bow shock region—the shock wave in the solar wind that surrounds the magnetosphere—is shown in blue. (Credit: European Space Agency)

Originally slated for a 2008 decommission, Cassini-Huygens's mission was extended until 2017. Cassini found that Saturn's rings are a complex place where small moons and ring particles jostle and collide, and where waves and jets constantly form and dissipate. Cassini discovered an icy plume shooting from Enceladus, and subsequent observations have revealed that the spray contains complex organic chemicals. Questions surrounding Enceladus's astrobiological potential are at the heart of many investigations that are being conducted on the mission. In addition, Cassini investigated the structure and complex organic chemistry of Titan's thick, smog-filled atmosphere. On the frigid surface, the spacecraft and its Huygens probe revealed vast methane lakes and widespread stretches of wind-sculpted hydrocarbon sand dunes. Cassini researchers also deduced the presence of an internal, liquid water-ammonia ocean.

For more information on the Huygens Probe and the Cassini-Huygens mission, please visit *http://www.esa.int/SPECIALS/Cassini-Huygens*, *http://saturn.jpl.nasa.gov/*, and *http://www.nasa.gov/cassini*.

Europe

Herschel

The European Space Agency's (ESA) Herschel Space Observatory is a space-based telescope that uses the light of the far-infrared and submillimeter portions of the spectrum to study the universe. It is able to reveal new information about the earliest, most distant stars and galaxies, as well as those closer to Earth in space and time. Herschel contains the largest single mirror ever built for a space telescope.

The Herschel mission launched in May 2009 together with ESA's Planck spacecraft, and the two spacecraft shared the same 4-month journey to independently enter their orbits between the Sun and Earth. As expected, the liquid-helium coolant on board the Herschel spacecraft was exhausted in April 2013, ending the space-based portion of the mission; analysis of the data collected, however, will continue for many years to come.

The Herschel telescope focuses light onto three instruments: the Spectral and Photometric Imaging Receiver (SPIRE), the Heterodyne Instrument for the Far-Infrared (HIFI), and the Photoconductor Array Camera and Spectrometer (PACS). SPIRE has two main components: a low-to-medium-resolution spectrometer and a photometer. It detects protons directly through a web-like bolometer, an instrument that can detect very small amounts of energy and convert them into electrical signals. The United Kingdom provided the majority of the design, development, and calibration of the SPIRE instrument, in cooperation with NASA. HIFI is a very-high-resolution heterodyne spectrometer. Rather than producing pictures of stars and galaxies, HIFI provides extremely detailed spectra of their atoms and molecules. The Netherlands provided most of the support for the HIFI instrument, also in collaboration with NASA, which provided the NASA Herschel Science Center (NHSC) for data archiving and analysis.

For more information on the Herschel Space Observatory, please visit *http://herschel.jpl.nasa.gov/index.shtml* and *http://sci.esa.int/science-e/www/area/index.cfm?fareaid=16*.

An artist's concept of the Herschel Space Observatory. (Credit: European Space Agency)

Europe

Mars Express

Mars Express, named for its rapid and streamlined development, marks the European Space Agency's (ESA) first visit to another planet in the solar system. Mars Express is helping to answer fundamental questions about the geology, atmosphere, surface environment, history of water, and potential for life on Mars.

The ESA-led Mars Express mission, launched in 2003, consists of an orbiter with seven experiments and a lander called Beagle 2. The orbiter has been a huge scientific success since it entered orbit in December 2003. ESA lost contact with the Beagle 2 lander after it was released from the orbiter, but its loss has not discouraged ESA from planning future attempts to land on Mars.

An artist's concept of Mars Express orbiting the Red Planet. (Credit: European Space Agency)

Working with ESA, NASA provided scientific and engineering support to the Mars Express mission, as well as significant Deep Space Network communications coverage for the mission. NASA worked on a bilateral basis with several ESA member states on the Beagle 2 lander and each of the seven experiments on the orbiter. NASA's main hardware contributions for the orbiter involved the joint development of the Mars Advanced Radar for Subsurface and Ionospheric Sounding (MARSIS) instrument with the Italian Space Agency (ASI) and the Analyzer of Space Plasmas and Energetic Atoms version 3 (ASPERA-3) instrument with the Swedish Institute of Space Physics (IRF). NASA also contributed to the development of the following instruments: the high-resolution stereo camera (HRSC) and the Mars Radio Science Experiment (MaRS) provided by the German Aerospace Center (DLR), the ASI-provided Planetary Fourier Spectrometer (PFS), and the Visible and Infrared Mineralogical Mapping Spectrometer (OMEGA) and Spectroscopy for Investigation of Characteristics of the Atmosphere of Mars (SPICAM) provided by the French National Center for Space Studies (CNES).

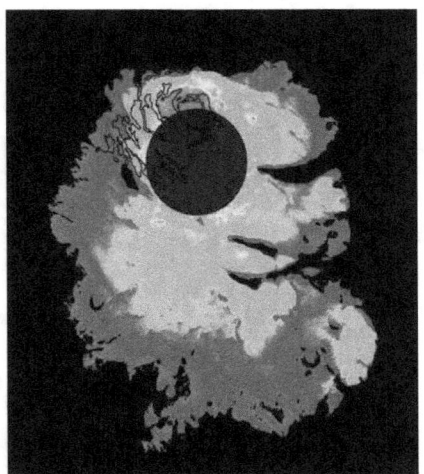

This map shows the thickness of layered deposits at the south polar region of Mars, an ice-rich geologic unit that was probed by the MARSIS instrument.

Mars Express has revolutionized scientists' view of the Red Planet. The MARSIS instrument made new measurements of Mars's south polar region that may indicate the presence of extensive frozen water. MARSIS also found evidence of buried impact basins. The SPICAM instrument discovered an aurora around Mars, and the OMEGA instrument revealed that the polar caps on Mars consist chiefly of carbon dioxide (CO_2) and exhibit seasonal variations in their coverage.

The OMEGA spectrometer on the Mars Express orbiter has aided in the discovery of clay materials, indicating that the planet once hosted warm, wet conditions. Concerning these recent discoveries, scientists have formulated a hypothesis that suggests Martian environments with abundant liquid water on the surface existed only for short episodes of time.

For more information on ESA's Mars Express mission, please visit *http://www.esa.int/SPECIALS/Mars_Express/index.html* and *http://mars.jpl.nasa.gov/express/*.

Science

Europe

Planck

From its position at the L2 Lagrange point, the European Space Agency (ESA)–led Planck mission, launched in 2009, is able to measure the cosmic microwave background (CMB) across nearly the full sky. Planck measures both minute temperature fluctuations and the polarization of the CMB, which is leading to a better understanding of the large-scale structure and history of the universe. With its precision instruments, Planck extracts essentially all of the cosmological information contained within the temperature map of the CMB. In addition, its map of the CMB polarization provides further insights into the evolution of the universe.

The Planck spacecraft consists of a telescope and two instruments—the high-frequency instrument (HFI) and the low-frequency instrument (LFI)—that measure neighboring bands of the CMB spectrum. The French National Center for Space Studies (CNES), NASA, and scientific institutes in the United Kingdom were the primary collaborators on the design, construction, and integration of the HFI. The Italian Space Agency (ASI), NASA, scientific institutes in the United Kingdom, and the Finnish Funding Agency for Technology and Innovation (Tekes) were the principal parties in the construction and integration of the LFI. In addition to executing the launch, ESA provided the telescope and spacecraft for the Planck mission.

For more information on the Planck mission, please visit *http://sci.esa.int/planck*.

An artist's concept of the Planck spacecraft. (Credit: European Space Agency)

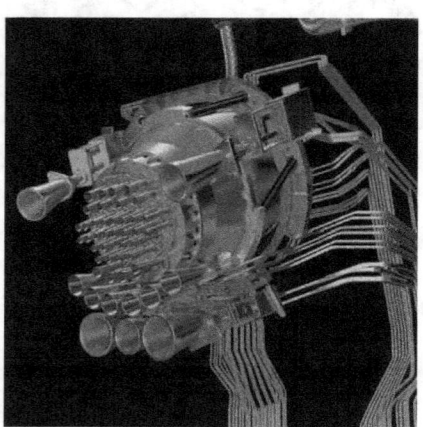

Located in the focal plane of the telescope, Planck's low-frequency instrument and high-frequency instrument are equipped with a total of 74 detectors covering nine frequency channels. Microwave light collected by the telescope reaches the instrument detectors via the conical feed horns. (Credit: European Space Agency)

Europe

Solar Orbiter

Solar Orbiter is a European Space Agency (ESA) mission with significant NASA participation that will explore the near-Sun environment to improve our understanding of how the Sun creates the environment of the inner solar system and, more broadly, generates the heliosphere itself, as well as how fundamental plasma physical processes operate near the Sun. NASA is contributing two instruments to the mission and is also providing the launch of the spacecraft. ESA is providing the spacecraft bus, integration of the instruments onto the bus, mission operations, and overall science operations. Individual European nations are providing the remainder of the ten-instrument payload. The Solar Orbiter mission is scheduled to be launched in 2017 from Cape Canaveral Air Force Station (CCAFS) in Florida.

Solar Orbiter will provide close-up views of the Sun's polar regions and its far side, and it will tune its orbit to the direction of the Sun's rotation to allow the spacecraft to observe a specific region for much longer than is currently possible. This will provide better insight into the evolution of sunspots, active regions, coronal holes, and other solar features and phenomena.

NASA's two instrument contributions are the Solar Orbiter Heliospheric Imager (SoloHI), a wide-field heliospheric imager provided by the Naval Research Laboratory; and a Heavy Ion Spectrometer (HIS) provided by Southwest Research Institute, part of the Solar Wind Plasma Analyzer (SWA). NASA is also collaborating individually with Belgium, France, Italy, Switzerland, and the United Kingdom on these Solar Orbiter instruments.

For more information on Solar Orbiter, please visit *http://science.nasa.gov/missions/solar-orbiter/*.

An artist's concept of Solar Orbiter observing the Sun. (Credit: European Space Agency)

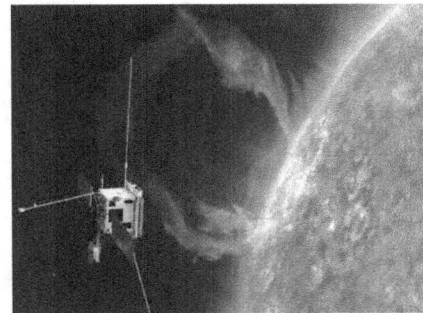

Solar Orbiter, ESA's next-generation Sun explorer, will investigate the connections and the coupling between the Sun and the heliosphere, a huge bubble in space created by the solar wind. The solar wind can cause auroras and disrupt satellite-based communication. (Credit: European Space Agency)

Science

An artist's rendering of a coronal mass ejection sweeping past one of the STEREO satellite observatories.

The far side unveiled! This is the first complete image of the solar far side, the half of the Sun invisible from Earth. Captured in June 2011, this composite image was assembled from NASA's two STEREO spacecraft.

Europe

STEREO
Solar TErrestrial RElations Observatory

By flying two spacecraft in orbit around the Sun while separated such that each has a view of the opposite sides of our closest star, NASA's Solar TErrestrial RElations Observatory (STEREO) mission provides a unique and revolutionary view of the Sun-Earth system.

The French National Center for Space Studies (CNES), German Aerospace Center (DLR), Hungarian Space Office (HSO), Swiss Space Office (SSO), and the United Kingdom (U.K.) Space Agency are all providing support for this mission, which NASA launched in 2006.

The two nearly identical spacecraft—one ahead of Earth in its solar orbit, the other trailing behind—have traced the flow of energy and matter from the Sun to Earth. STEREO has also revealed the three-dimensional (3-D) structure of coronal mass ejections (CMEs). CMEs travel away from the Sun at speeds of approximately one million miles per hour (mph) and can create major disturbances in the interplanetary medium and trigger severe magnetic storms when they collide with Earth's magnetosphere. The STEREO satellites are a key addition to the fleet of space weather detection satellites because, by taking advantage of their unique side-viewing perspective, they provide more accurate alerts for the arrival time of Earth-directed solar ejections.

Each STEREO spacecraft has four suites of instruments—three led by U.S. scientists and one led by French scientists. The STEREO/Wind/Radio and Plasma Wave Experiment (S/WAVES) is an interplanetary radio-burst tracker that traces the generation and evolution of traveling radio disturbances from the Sun to the orbit of Earth. CNES is sponsoring the principal investigator and provided this instrument suite. In-situ Measurements of Particles and CME Transients (IMPACT) samples the 3-D distribution of these solar energetic particles, provides the plasma characteristics of them, and measures the local vector magnetic field. The European Space Agency (ESA), CNES, DLR, HSO, and SSO contributed to IMPACT. Sun Earth Connection Coronal and Heliospheric Investigation (SECCHI) consists of an extreme-ultraviolet imager, two white-light coronagraphs, and a heliospheric imager. SECCHI studies the 3-D evolution of CMEs from birth at the Sun's surface through the corona and interplanetary medium to its eventual impact at Earth. CNES, DLR, and the U.K. Space Agency contributed to SECCHI. The PLAsma and SupraThermal Ion Composition (PLASTIC) instrument provides key diagnostic measurements on the form of mass and charge-state composition of heavy ions and characterizes the CME plasma from ambient coronal plasma. DLR, HSO, and SSO contributed to PLASTIC.

For more information on STEREO, please visit *http://www.nasa.gov/stereo*.

European Space Agency

Cluster-II

Since being launched into an elliptical polar orbit in the summer of 2000, the European Space Agency's (ESA) four identical Cluster-II satellites have continuously crossed from Earth's magnetic field into the solar wind and back again, resulting in the most detailed investigation ever of the interaction between the Sun and Earth. More than a dozen years later, the Cluster-II quartet has provided the first three-dimensional (3-D) view of near-Earth space and continues to unravel the secrets of the invisible particles and magnetic fields that envelop our planet.

Cluster-II's four satellites—Rumba, Samba, Salsa, and Tango—fly in formation around Earth to provide a 3-D picture of how the continuous solar wind, composed of charged particles and plasma from the Sun, affects the near-Earth space environment and its protective magnetic bubble, known as the magnetosphere. Each spacecraft carries an identical suite of 11 scientific instruments, and the satellites are positioned in orbit to form the four points of a pyramid. This arrangement allows for the collection of data that can be used to generate 3-D views of structures in both the magnetosphere and solar wind.

An artist's concept of the Cluster-II satellites. (Credit: European Space Agency)

As a cooperative science project, ESA and NASA have worked closely together throughout the mission. ESA is providing the management of the satellites and the science operations, while NASA-sponsored investigators are participating in the operation of the scientific instruments, data processing, and analyses of the Cluster-II scientific data.

An illustration of solar wind arriving at Earth's magnetosphere.

During its years of discovery, Cluster-II has provided numerous new insights into the fundamental processes that influence not only near-Earth space, but also the universe as a whole. Cluster-II data has been used to study magnetic reconnection, a physical process that can cause the ejection of solar plasma jets toward Earth that affect Global Positioning System (GPS) and radio communications. Cluster-II has confirmed that black auroras—strange electrical phenomena that generate dark, empty regions within the Northern and Southern Lights—are a kind of "anti-aurora" that draw electrons from the ionosphere. Cluster-II was also able to make a key discovery about "killer electrons," which are high-energy particles trapped in the outer Van Allen radiation belt. These particles move close to the speed of light and carry enough energy to penetrate satellite shielding and cause electrical shorts that can seriously damage vital components.

For more information on the Cluster-II mission, please visit *http://sci.esa.int/science-e/www/area/index.cfm?fareaid=8*.

Science

European Space Agency

Euclid

An artist's impression of Euclid. (Credit: European Space Agency)

Euclid is a European Space Agency (ESA)–led mission to map the geometry of the dark universe. Seeking to understand the origin of the universe's accelerating expansion, Euclid will probe dark energy and measure cosmological parameters. It will employ cosmological probes to investigate the nature of dark energy, gravity, and dark matter by tracking their observational signatures in the geometry of the universe and the cosmic history of structure formation. ESA is scheduled to launch Euclid in 2019, and the mission is designed for a 6-year lifespan.

Euclid will use two complementary techniques to accomplish its mission: weak gravitational lensing (WL) and Baryonic Acoustic Oscillations (BAO). WL is a unique technique that maps dark matter and measures dark energy by quantifying the apparent distortions of galaxy images, thereby detecting changes in the galaxy's observed ellipticity that are caused by mass inhomogeneities along the line of sight. The lensing signal is derived from the statistical measurement of the slight distortions in the shapes of galaxies, taking into account their distances. In addition, Euclid surveys yield data on several important complementary cosmological phenomena such as galaxy clustering, redshift space distortions, and the integrated Sachs-Wolfe effect.

Euclid consists of a 1.2-meter Korsch telescope that directs light to two instruments: the visual imager (VIS) and the near-infrared spectrograph and photometer (NISP). NASA will provide near-infrared (NIR) detectors for NISP. VIS provides high-quality images to carry out the weak-lensing galaxy-shear measurements. NISP performs imaging photometry to provide near-infrared photometric measurements for photometric redshifts, and it also carries out slitless spectroscopy to obtain spectroscopic redshifts.

Euclid's surveys will yield unique data in various fields of astrophysics and will provide a primary database for next-generation multiwavelength surveys. The mission is planned to produce a dataset, with images and photometry, of more than a billion galaxies and several million spectra.

For more information on Euclid, please visit *http://euclid.gsfc.nasa.gov/*.

European Space Agency

ExoMars

ExoMars is the flagship mission set of the European Space Agency's (ESA) robotic exploration program. Consisting of missions to Mars in 2016 and 2018, the ExoMars program seeks to establish whether life ever existed on Mars—one of the outstanding scientific questions of our time. In its investigation of the Martian environment, the ExoMars program will demonstrate new technologies, thereby paving the way for a future Mars Sample Return mission in the 2020s. In 2012, Russia joined ESA as a partner on these missions.

Two missions are foreseen within the ExoMars program: one consisting of an orbiter and an Entry, Descent, and Landing (EDL) Demonstrator Module (EDM), to be launched in 2016; and the other, with a launch date of 2018, featuring a rover to be landed on the surface of Mars. The ExoMars program will demonstrate a number of essential flight and in situ enabling technologies that are necessary for future exploration missions, such as an international Mars Sample Return mission, including: EDL of a payload on the surface of Mars; surface mobility with a rover; access to the subsurface to acquire samples; and sample acquisition, preparation, distribution, and analysis. At the same time, a number of important scientific investigations will be carried out, including: searching for signs of past and present life on Mars, investigating how the water and geochemical environment varies, and investigating Martian atmospheric trace gases and their sources.

An artist's concept of the ExoMars orbiter. (Credit: European Space Agency)

The 2016 mission will include the ExoMars Trace Gas Orbiter, which will carry the EDM as well as its own suite of scientific instruments to detect and study atmospheric trace gases, such as methane. The EDM will contain sensors to evaluate the lander's performance as it descends, and additional sensors to study the environment at the landing site. NASA is planning to provide the Electra telecommunications package on the orbiter, support science co-investigators on some of the orbiter's instruments, provide engineering support to the mission as a whole, and provide data-relay support to the EDM on a best-efforts basis.

The 2018 mission will center on the ExoMars rover, which will carry a drill and a suite of instruments dedicated to exobiology and geochemistry research. NASA intends to provide the mission with engineering support and to work with the German Aerospace Center (DLR) in providing portions of the Mars Organic Molecule Analyzer (MOMA) instrument.

International cooperation is recognized as a cornerstone of future Mars exploration, and NASA and ESA are continuing a dialogue about possible future opportunities to collaborate beyond the 2018 launch opportunity.

For more information on ESA's ExoMars program, please visit *http://exploration.esa.int/science-e/www/area/index.cfm?fareaid=118*.

Science

European Space Agency

HST
Hubble Space Telescope

The European Space Agency (ESA) joined NASA in developing the Hubble Space Telescope (HST) in 1977. HST was launched by the Space Shuttle into low-Earth orbit in 1990. ESA provided the Faint Object Camera (FOC), an instrument for high-resolution imagery in the ultraviolet, visual, and near-infrared portions of the spectrum; two 12-meter solar arrays to provide power for HST; and a second set of solar arrays to replace the first set. This second set of arrays was installed during Servicing Mission 1 in December 1993, and it operated for over 8 years. The second set of solar arrays was replaced with more robust solar arrays during Servicing Mission 3B in March 2002.

ESA was represented on the HST Science Working Group, which was the principal mechanism for scientific input to the HST Project Office during the development of HST. ESA provided post-launch operational support, with experts located at the HST project office and the science operations facility. European astronomers receive approximately 15 percent of HST viewing time.

Over its 22-year history, HST has been used by astronomers to: take over a million exposures and assemble hundreds of thousands of images that have expanded our knowledge of the universe; take images of galaxies in the local universe such as Messier 101 (M101), capturing the largest and most detailed photo of a spiral galaxy ever released; complete an 8-year effort to measure the expansion of the universe, which resulted in one of the most important scientific discoveries of our time—that the universe is expanding at an accelerating rate, driven by an unknown force; provide convincing evidence for the existence of black holes several billion times the mass of the Sun in the nuclei of some giant galaxies, as well as find evidence that supermassive black holes are at the core of most, if not all, galaxies; discover 16 extrasolar planet candidates orbiting a variety of distant stars in the central region of our Milky Way galaxy; and confirm the presence of two moons around the distant minor planet Pluto.

For more information on HST, please visit *http://hubble.nasa.gov/* and *http://www.spacetelescope.org/about/index.html*.

This photo of the HST was taken from Space Shuttle Columbia during the STS-109 mission.

This detailed picture of the Helix Nebula shows a fine web of filaments, like the spokes of a bicycle, embedded in the colorful red and blue gas ring around a dying star. The Helix Nebula is one of the nearest planetary nebulae to Earth—only 650 light years away. (Credit: NASA, National Optical Astronomy Observatory, ESA, the Hubble Helix Nebula Team, M. Meixner [Space Telescope Science Institute], and T.A. Rector [National Radio Astronomy Observatory])

European Space Agency

ISS
International Space Station

The European Space Agency's (ESA) primary contributions to the International Space Station (ISS) Program are the Columbus Laboratory, Nodes 2 and 3, the Cupola, the Automated Transfer Vehicle (ATV), and the European Robotic Arm (ERA).

The ISS Program is a partnership among the space agencies of Canada, Europe, Japan, Russia, and the United States. International crews have continuously inhabited the ISS since November 2000. Cargo transportation is provided by European, Russian, and Japanese vehicles and by NASA's contracts for Commercial Resupply Services (CRS). Crew transportation and rescue is currently provided by the Russian Federal Space Agency and, beginning in 2017, will be provided by NASA's commercial crew transportation services.

NASA's contributions to the ISS Program include overall management, coordination, and integration of the ISS Program across the partnership for operations, safety, transportation, and integrated hardware performance. NASA hardware components include the U.S. laboratory module, the airlock for extravehicular activities (EVA), truss segments to support the U.S. solar arrays and thermal radiators, three connecting nodes, and living quarters.

The ISS functions as an orbital microgravity and life sciences laboratory, a test bed for new technologies in areas such as life support and robotics, and a platform for astronomical and Earth observations. The ISS also serves as a unique engineering test bed for flight systems and operations critical to NASA's future exploration missions. U.S. research on the ISS concentrates on the long-term effects of space travel on humans and on technology development activities in support of exploration. The ISS is a stepping-stone for human exploration and scientific discovery beyond low-Earth orbit.

ESA's Columbus Laboratory houses ISS research facilities for experiments in life and materials science, fluid physics, biosciences, and technological applications. Node 2, built by ESA for NASA, connects the U.S., European, and Japanese laboratories and provides utilities essential to support life on the ISS. The Cupola and Node 3 were both built for NASA by ESA. The Cupola is a pressurized observation and control tower for the ISS with windows that provide a panoramic view through which crew members can observe and guide operations on the outside of the ISS. Node 3 is the central home for ISS's environmental control equipment. ESA's ATVs are logistical resupply vehicles that dock to the ISS and provide dry cargo, atmospheric gas, water, and propellant delivery. ESA is providing five ATVs for the ISS. The European Robotic Arm is based on the Russian Multipurpose Laboratory Module and can be moved to different locations on the exterior of the ISS.

For more information on the ISS and ESA's contributions, please visit *http://www.esa.int/esaHS/ESAQEI0VMOC_iss_0.html* and *http://www.nasa.gov/mission_pages/station/main/index.html*.

This photo of the European Space Agency's Columbus Laboratory was taken by the Space Shuttle Endeavour's crew on STS-127 as the orbiter circled the Space Station.

The European Space Agency's "Edoardo Amaldi" Automated Transfer Vehicle-3 (ATV-3) approaches the International Space Station in March 2012. With a successful docking, the unmanned cargo spacecraft delivers 220 pounds of oxygen, 628 pounds of water, 4.5 tons of propellant, and nearly 2.5 tons of dry cargo, including experiment hardware, spare parts, food, and clothing.

Human Exploration and Operations

European Space Agency

ISS Early Utilization

Recognizing the mutual NASA and European Space Agency (ESA) interest in International Space Station (ISS) utilization during early phases of the program, the agencies initiated cooperative activities to ensure early research opportunities. This cooperation permitted the location and operation of hardware and experiments in the U.S. Laboratory Module (Destiny) prior to the delivery of the European Columbus Laboratory. It also provided flight opportunities for two ESA astronauts. As part of this cooperation, ESA contributed a broad range of hardware, including the Microgravity Science Glovebox (MSG), Minus-80°C Laboratory Freezer for ISS (MELFI), and the European Modular Cultivation System (EMCS).

The MSG is a sealed container with built-in gloves that provides an enclosed workspace for investigations conducted in the microgravity environment of the ISS. It was launched on Space Transportation System (STS)–111 in 2002 and is currently housed in the Columbus Laboratory. It provides a safe environment for research with liquids, flames, and particles used as a part of everyday research in ground-based facilities on Earth. Without the MSG, many types of hands-on investigations on the ISS would be impossible or severely restricted.

The MELFI was launched on STS-121 in 2006. It is a rack-sized facility designed to provide the ISS with refrigerated volume for the storage and fast freezing of life science and biological samples.

The EMCS was also launched on STS-121. It is a combination centrifuge and growth chamber that is designed to carry out plant-growth experiments in controlled partial gravity and microgravity conditions. EMCS experiments contribute to food production research that may benefit astronauts on long-term missions, such as an expedition to Mars. These experiments are used for research on insects, amphibians, and invertebrates, as well as for studies with cell and tissue cultures. Furthermore, plant-cell-root and physiology experiments provide insight into the growth processes in plants, which could help to improve food-production techniques both in space and on Earth.

Materials Science Research Rack–1 (MSRR-1) was delivered to the ISS during STS-128 in 2009. It is the primary facility for U.S.-sponsored materials science research on the ISS and is contained in an International Standard Payload Rack (ISPR) that is equipped with the Active Rack Isolation System (ARIS) for the best possible microgravity environment. The MSRR-1 accommodates dual experiment modules and provides simultaneous on-orbit processing operations capability. The first integral experiment module for the MSRR-1, the Materials Science Laboratory, is also a cooperative activity between NASA and ESA.

For more information, please visit *http://www.nasa.gov/mission_pages/station/research/index.html*.

European Space Agency astronaut Paolo Nespoli, Expedition 27's flight engineer, works with the Microgravity Science Glovebox in the Destiny Laboratory of the International Space Station.

In the European Modular Cultivation System experiment container with Tropi Experiment equipment, plants that sprouted from seeds were videotaped, and samples were collected for analysis at a molecular level to determine what genes are responsible for successful plant growth in microgravity.

European Space Agency

ISS HRF-EPM
International Space Station Human Research Facility–European Physiology Module

The European Space Agency (ESA) and NASA are cooperating on the ESA European Physiology Module (EPM) and the NASA Human Research Facility (HRF). ESA provided the EPM, a single-rack multi-user facility launched on Space Transportation System (STS)–122 in 2008 for the Columbus Laboratory on the International Space Station (ISS). The Columbus Laboratory module is ESA's most significant contribution to the ISS.

An interior view of the European Columbus Laboratory at the Space Station Processing Facility at NASA's Kennedy Space Center in Florida as the module is prepared for flight to the International Space Station: European Drawer Rack (left), European Physiology Modules Facility (center), Biolab and Fluid Science Lab in the launch configuration (right). (Credit: European Space Agency/S. Corvaja)

The EPM and other components support neuroscience research and investigations of respiratory and cardiovascular conditions, hormonal and body fluid shift, and bone demineralization. The experiment results also contribute to researchers' understanding of terrestrial health problems, such as the aging process, osteoporosis, balance disorders, and muscle wastage.

Built in Bremen, Germany, the EPM includes the following containers for science experiments: ESA's Multi-Electrode Electroencephalogram Mapping Module (MEEMM), for neurologic scans of the brain; ESA's Samples Collection Kit, for collection of different medical instruments and boxes for biologic probes; and the German/French Cardiolab, for scanning of the cardiovascular system.

European Space Agency astronaut Thomas Reiter, Expedition 13's flight engineer, prepares the HRF-2 rack for an upgrade of the Pulmonary Function System experiment in the Destiny Laboratory of the International Space Station.

In conjunction with the EPM, NASA's HRF supports the science community by providing the instrumentation and capabilities needed to collect and distribute data gained through human life-science research in space. NASA and ESA reviewed each other's plans for the development of research hardware to minimize duplication and reduce costs while establishing a research capability that will meet the requirements of both agencies for biomedical research.

The HRF provides a space for experiments and instruments, along with the means to access ISS services and utilities such as electrical power, command and data handling, air and water cooling, and pressurized gases.

For more information on the HRF and EPM, please visit *http://www.esa.int/esaHS/ESA44I0VMOC_iss_0.html*, *http://www.esa.int/esaHS/ESAFRG0VMOC_iss_0.html*, and *http://www.nasa.gov/mission_pages/station/research/experiments/EPM.html*.

Science

European Space Agency

JWST
James Webb Space Telescope

The James Webb Space Telescope (JWST), planned for launch in 2018, will be a large infrared telescope with a 6.5-meter primary mirror. JWST will enable the study of every phase in the history of our universe, ranging from the first luminous glows after the Big Bang, to the formation of solar systems capable of supporting life on Earth-like planets, to the evolution of our own solar system. JWST is an international collaboration between NASA, the Canadian Space Agency (CSA), and the European Space Agency (ESA).

Referred to as the JWST Observatory, the space-based portion of the JWST system consists of three parts: the Optical Telescope Element (OTE), the Spacecraft Element, and the Integrated Science Instrument Module (ISIM). The OTE is the eye of the JWST, collecting light from space and providing it to scientific instruments. The Spacecraft Element is composed of the spacecraft bus (which provides support functions for the observatory) and the sunshield system (which separates the observatory into a warm Sun-facing side and a cold side that faces away from the Sun). The ISIM contains four instruments: the Mid-Infrared Instrument (MIRI), the Near-Infrared Spectrograph (NIRSpec), the Near-Infrared Camera (NIRCam), and the Fine Guidance Sensor/Near-Infrared Imager and Slitless Spectrograph (FGS/NIRISS). JWST's instruments will be designed to work primarily in the infrared range of the electromagnetic spectrum, with some capability in the visible range.

Several innovative technologies have been developed for JWST, including a folding, segmented primary mirror that is adjusted to shape after launch; ultra-lightweight beryllium optics; detectors capable of recording extremely weak signals; microshutters that enable programmable object selection for the spectrograph; and a cryocooler for cooling the mid-infrared detectors to the required 7° kelvin operating temperature.

ESA is providing the Ariane 5 Launch Vehicle on which JWST will be launched. ESA is also providing two science instruments: the NIRSpec and the MIRI Optical Bench Assembly (OBA). NASA is responsible for the design and development of the JWST Observatory and for overall management of the mission.

For more information on JWST, please visit *http://jwst.gsfc.nasa.gov/*.

An artist's concept of the JWST.

Designed to block solar light and keep the JWST Observatory operating at cryogenic temperature, the five-layer sunshield consists of thin membranes made from a polymer-based film and supporting equipment such as spreader bars, booms, cabling, and containment shells.

European Space Agency

JUICE
JUpiter ICy moons Explorer

The European Space Agency's (ESA) JUpiter ICy moons Explorer (JUICE) mission, the first large-class mission in ESA's Cosmic Vision 2015–2025 program, will spend at least 3 years making detailed observations of the biggest planet in the solar system and three of its largest moons: Ganymede, Callisto, and Europa. Planned for launch in 2022 and arrival at Jupiter in 2030, JUICE will study these moons—which are thought to harbor vast water oceans beneath their icy surfaces—and map their surfaces, sound their interiors, and assess their potential for hosting life in their oceans.

In 2013, ESA selected the JUICE payload complement of 10 instruments plus an experiment using the Very Long Baseline Interferometry (VLBI) technique, which will be developed by scientific teams from 16 European countries, the United States, and Japan. The payload complement will address the mission's science objectives, from in situ measurements of Jupiter's atmosphere and plasma environment to remote observations of the surface and interior of the three icy moons. NASA will provide the ultraviolet spectrometer (UVS), and it will also contribute to the Italian Space Agency (ASI)–led Radar for Icy Moon Exploration (RIME) and the Swedish Institute of Space Physics–led Particle Environment Package (PEP).

An artist's concept of JUICE orbiting Jupiter and exploring Ganymede. Ganymede is the only moon in the Solar System known to generate its own magnetic field, and JUICE will seek to observe its unique magnetic and plasma interactions with Jupiter's magnetosphere in detail. (Credit: European Space Agency)

The NASA-provided UVS, whose mission and science team is led by the Southwest Research Institute, will use image capture to explore the surfaces and atmospheres of Jupiter's icy moons and how they interact with the Jupiter environment. The instrument will also determine how Jupiter's upper atmosphere interacts with the lower atmosphere below it and the ionosphere and magnetosphere above it. The instrument will also provide images of the aurora on Jupiter and Ganymede. The Johns Hopkins University Applied Physics Laboratory (APL) is leading the NASA contribution to PEP, which will include instruments to measure the neutral material and plasma that are accelerated and heated to extreme levels in Jupiter's fierce and complex magnetic environment. The Jet Propulsion Laboratory (JPL) is leading NASA's RIME contribution: a transmitter and receiver hardware for a radar sounder designed to penetrate the icy crust of Europa, Ganymede, and Callisto to a depth of about 9 kilometers. This will allow scientists to see for the first time the underground structure of these tectonically complex and unique icy worlds.

For more information on the JUICE mission, please visit *http://sci.esa.int/juice*.

European Space Agency

LISA Pathfinder
Laser Interferometer Space Antenna Pathfinder

The Laser Interferometer Space Antenna (LISA) Pathfinder mission, scheduled to launch in 2014, is a collaborative mission between NASA and the European Space Agency (ESA) to demonstrate technologies associated with future gravitational-wave observatory concepts. The LISA Pathfinder mission aims to validate the performance of technologies for precise spacecraft control.

The LISA Pathfinder mission consists of two test packages: the LISA Technology Package (LTP) provided by ESA and the Disturbance Reduction System (DRS) provided by NASA. The LTP is designed to comprise one gravitational sensor, an optical bench, and a related optical meteorology system. The LTP sensors are used to monitor the position of the spacecraft with respect to a freely floating test mass. The DRS is designed to include micro-propulsion thrusters, supporting electronics, housekeeping avionics, data handling, and drag-free attitude control software. The thrusters will be used to adjust the position of the spacecraft to keep it centered on the test mass. Both of these test packages lay important groundwork for potential future space-based gravitational-wave observatories. For this ESA-led mission, in addition to providing the LTP, ESA will construct the LISA Pathfinder spacecraft and operate the instruments after launch.

For more information on the LISA Pathfinder mission, please visit *http://nmp.jpl.nasa.gov/st7/ABOUT/About_index.html* and *http://sci.esa.int/science-e/www/area/index.cfm?fareaid=40*.

An artist's rendering of the LISA Pathfinder spacecraft and its propulsion module after separation. (Credit: European Space Agency)

An artist's rendering of the LISA Technology Package flying on board the LISA Pathfinder spacecraft. The partly transparent view reveals the proof masses: two 46-mm cubes of a gold-platinum alloy that are housed in individual vacuum cans. The cubes serve both as mirrors for the laser interferometer (red light paths) and as inertial references for the spacecraft's drag-free control system. (Credit: European Space Agency)

European Space Agency

Rosetta

The European Space Agency's (ESA) Rosetta spacecraft will be the first to explore a comet at close quarters over an extended period of time. The mission is comprised of a large orbiter, built to operate for a decade at great distances from the Sun, and a small lander called Philae. The orbiter and lander each carry a sizable complement of scientific experiments designed to carry out the most detailed study of a comet ever attempted.

In 2004, ESA launched Rosetta on a 10-year journey to reach Comet 67P/Churyumov-Gerasimenko. The journey includes gravity assists from Earth and Mars and flybys of main belt asteroids. After entering orbit around Comet 67P/Churyumov-Gerasimenko in 2014, the orbiter will release its small lander onto the icy nucleus and then spend the next 2 years orbiting the comet as it heads toward the Sun.

NASA is providing mission operations and Deep Space Network communications support to the Rosetta mission. It is also contributing a number of instruments, including the Alice Ultraviolet Imaging Spectrometer, the Microwave Instrument for the Rosetta Orbiter (MIRO), the Rosetta Plasma Consortium (RPC) Ion and Electron Sensor (IES), and the Rosetta Orbiter Spectrometer for Ion and Neutral Analysis (ROSINA).

The Rosetta spacecraft participated in the observation campaign of NASA's Deep Impact encounter with Comet Tempel 1 on July 4, 2005. Rosetta has also conducted a co-observing session of the planet Jupiter with NASA's New Horizons mission.

For more information on ESA's Rosetta mission, please visit *http://rosetta.jpl.nasa.gov/*.

An artist's concept of the Rosetta spacecraft on its journey to the Churyumov-Gerasimenko comet. (Credit: European Space Agency)

An artist's concept of the small Philae lander on the surface of the Churyumov-Gerasimenko comet. (Credit: European Space Agency)

Science

European Space Agency

SOHO
Solar and Heliospheric Observatory

The Solar and Heliospheric Observatory (SOHO) is a joint project between the European Space Agency (ESA) and NASA to study the Sun from its deep core to its outer corona and solar winds. Since SOHO's launch in 1995, the observatory has been providing a vast amount of images and data regarding the Sun and the heliosphere. Originally designed for a 2-year mission, SOHO has been extended numerous times through 2016 to cover an entire 11-year solar cycle and the start of the new cycle.

This composite image put together from SOHO data shows a fiery coronal mass ejection with stunning, bright details in the ejected material.

The twelve instruments on board SOHO were provided by European and American teams of scientists who continue to operate the instruments and process the data. The instruments take images of the solar surface, the dynamic layers of the solar atmosphere, and the corona, which is observable only in ultraviolet light. By special processing of the solar white-light data, scientists can measure the magnetic fields around sunspots and look deep into the interior of the Sun. Other instruments measure the pulse and total light output of the Sun and the speed, density, and composition of solar wind as it flows past SOHO. ESA managed the construction, integration, and testing of the SOHO spacecraft, and a team of ESA scientists has been resident at NASA's Goddard Space Flight Center (GSFC) to orchestrate the planning of science observations. NASA provided the launch of SOHO, operates the spacecraft, and tracks it with the Deep Space Network.

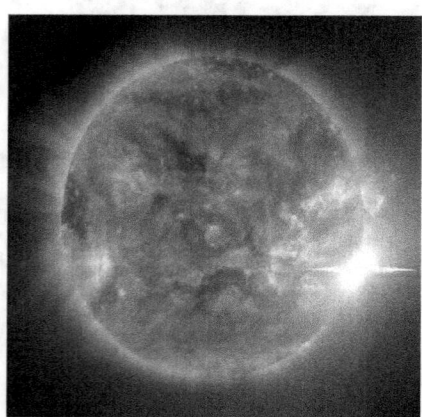

SOHO's Extreme-ultraviolet Imaging Telescope (EIT) captured this image of the Sun as it unleashed a powerful flare.

From its position orbiting around the L1 Lagrange point (about 1.5 million kilometers from Earth), SOHO has provided an unprecedented breadth and depth of information about the Sun, from its interior through its hot and dynamic atmosphere to the solar wind and its interaction with the interstellar medium.

Besides watching the Sun, SOHO has become the most prolific discoverer of comets in astronomical history. As of January 2011, amateur astronomers around the world have discovered more than 2,000 comets using images from SOHO's Large Angle and Spectrometric Coronagraph Experiment (LASCO) instrument.

For more information on the SOHO mission, please visit *http://sohowww.nascom.nasa.gov/*.

European Space Agency

XMM-Newton
X-ray Multi-Mirror-Newton

The European Space Agency's (ESA) X-ray Multi-Mirror (XMM)–Newton satellite, launched into Earth orbit in 1999, carries three advanced x-ray telescopes. Each of XMM-Newton's three telescopes contains 58 high-precision concentric mirrors, which are nested to offer the largest collecting area possible to catch x rays. Additionally, its detectors are very sensitive, enabling the telescopes to observe faint objects.

XMM-Newton uses an unusual technique to focus x rays onto the different detectors. An optical telescope works by reflecting and focusing light; however, x rays have such high energy that they pass through most materials, making reflection impossible. Therefore, in the case of an x-ray telescope, the mirrors are arranged so that incoming x rays graze off of the mirrors. The shape and highly polished surfaces of the mirrors focus the incoming x rays into a beam when they enter the detectors.

An artist's rendering of XMM-Newton orbiting Earth.
(Credit: European Space Agency)

The XMM-Newton spacecraft consists of three main instruments. The European Photon Imaging Camera (EPIC)—the main camera onboard XMM-Newton—can perform both x-ray imaging and spectroscopy. Actually three cameras in one, it employs three different detectors to view the x rays. The second instrument is a Reflection Grating Spectrometer (RGS), which is similar to an optical prism in that it spreads out the incoming light according to the wavelength of the photon. The third instrument is the Optical/Ultraviolet Monitor (OM), which has a 30-centimeter-wide mirror designed to focus optical and ultraviolet light coming from the same direction as the x rays observed by the other instruments on XMM-Newton. This process allows for the simultaneous observations of a target in the optical/ultraviolet and x-ray regions of the electromagnetic spectrum.

XMM-Newton is able to detect and study celestial x-ray sources that are otherwise impossible to view from the planet due to Earth's protective atmosphere. Current studies using XMM-Newton have focused on black hole observations due to the large amounts of x rays that surround these bodies. Data from XMM-Newton, as well as from NASA's Swift satellite, have revealed the youngest black hole known to exist in our cosmic neighborhood. The newly discovered young black hole could help scientists better understand how massive stars explode, which ones leave behind black holes, and the number of black holes in our galaxy.

ESA designed and built the XMM-Newton spacecraft, along with the three main instruments aboard the spacecraft. ESA also controls spacecraft operations. NASA provides access to the mission for U.S. astronomers through the Guest Observer facility at the Goddard Space Flight Center (GSFC).

For more information on XMM-Newton, please visit *http://xmm.gsfc.nasa.gov/* and *http://xmm.sonoma.edu/index.html*.

Education and Outreach

Finland

Angry Birds Space

Astronaut Don Pettit demonstrates microgravity on the International Space Station.

An image from *Angry Birds Space*, used by permission of Rovio Entertainment Ltd.

NASA is cooperating with Rovio Entertainment Ltd., a Finnish video-game development company, on its production of a space-related version of the video game *Angry Birds*. This cooperation facilitates better public understanding of NASA's programs and missions, aids the dissemination of NASA educational and outreach materials to new audiences, and provides an engaging education opportunity for audiences worldwide. NASA's Space Act Agreement with Rovio also ensures reasonable and accurate depictions of space physics and NASA missions. To further these objectives, NASA provides Rovio with various materials, such as audio, video, transcripts, and images.

On March 8, 2012, NASA released a video by NASA astronaut Don Pettit, who gave a demonstration of microgravity and its effect on the trajectories of objects on board the International Space Station (ISS). He explained that this scientific principle would be seen in the new version of the game *Angry Birds Space*, when the trajectory of the game's fictional birds are changed by the gravitational fields of nearby planets. Rovio released *Angry Birds Space* to the public soon after, on March 22, 2012.

The video by Don Pettit is accessible while playing the game and can also be viewed at *http://www.nasa.gov/microgravity/*.

For more information on NASA and Rovio's cooperation on *Angry Birds Space*, please visit *http://www.nasa.gov/microgravity/* and *http://www.youtube.com/watch?v=lxI1L1RiSJQ*.

France

Aircraft Icing Research

A key NASA aeronautics research interest involves understanding the phenomena of aircraft icing. NASA and the French Aerospace Lab (ONERA) began conducting cooperative computational fluid dynamics (CFD) modeling of ice accretions in 2006 using subscale and full-scale airfoil models. The objectives of the collaboration were to establish an aerodynamic database for a full-scale airfoil with high-fidelity ice contamination, to develop methodologies for accurate subscale iced-aerodynamic simulation, to improve iced-CFD methods, and to better understand iced aerodynamics. Two ice-accretion test campaigns were completed in the NASA Glenn Research Center (GRC) Icing Research Tunnel (IRT) in Cleveland, Ohio, and two iced-aerodynamic test campaigns were completed in the ONERA F1 pressurized, low-speed, large-scale wind tunnel in Fauga-Mauzac, France.

This post-flight image shows ice contamination as a result of encountering Supercooled Large Droplet (SLD) conditions near Parkersburg, West Virginia.

High-fidelity, artificial ice shapes were made from the ice accreted on both a full-scale and a subscale two-dimensional airfoil model in the NASA IRT facility. The full-scale ice accretions were then used in aerodynamic tests at the ONERA F1 facility. During the first F1 campaign, aerodynamic performance data were obtained across a range of the aerodynamically differing artificial ice shapes. The second campaign was devoted to detailed Particle Image Velocimetry (PIV) flow measurements of the airflow over the ice-contaminated airfoil. The subscale artificial ice shapes were used in aerodynamic tests at the University of Illinois's low-speed tunnel. The F1 tests were the first such tests of full-scale, high-fidelity iced aerodynamics and produced a large benchmark database. This database, in conjunction with the subscale aerodynamic database, facilitated the development of improved methods for scaled iced-aerodynamic testing.

A commuter transport engine undergoes testing in the Icing Research Tunnel at the NASA Glenn Research Center.

For more information on NASA's icing research, please visit *http://ntrs.nasa.gov/archive/nasa/casi.ntrs.nasa.gov/20110023761_2011025037.pdf*.

Aeronautics Research

Technician Donald Day adjusts a one-quarter-scale model before a test in the Basic Aerodynamics Research Tunnel. NASA's Langley Research Center aeronautics researchers are studying how to make landing gear quieter.

A test rig is assembled at NASA's Glenn Research Center to evaluate the noise reduction of a newly developed metallic foam liner.

France

Airframe Noise and Environmental Noise Mitigation

A major goal of both NASA and the French Aerospace Lab (ONERA) is to mitigate environmental noise of civil aircraft during takeoff and landing. Both agencies have devoted significant resources to developing numerical methods for predicting airframe noise, and they are collaborating on means to both identify and reduce the noise associated with aircraft engines and landing gear.

Aircraft engines generate a significant portion of aircraft noise, and acoustic liners mounted upstream and downstream of an engine's rotor fan are used extensively for the reduction of this noise. Advances in the prediction and measurement of the effectiveness of acoustic liners to reduce noise radiated from aircraft engines are important to the continued improvement of engine noise reduction capabilities. These advances require the development of advanced simulation techniques along with high-fidelity experimental data.

A less obvious source of landing and takeoff noise comes from the airframe. High-lift devices on the wing and aircraft landing gear are currently the largest airframe noise sources for civil transport. Computational studies to understand the complex flow around landing gear may lead to methods to reduce this source.

These studies, when integrated with experimental flight-test and wind-tunnel data, could lead to quieter, more efficient airframe components, which will become part of a new generation of ultra-quiet, low-emission, subsonic aircraft designs, and lead to reduced fuel consumption and quieter skies over our neighborhoods.

For more information on airframe noise and environmental noise mitigation, please visit *http://www.nas.nasa.gov/SC10/Khorrami_LandingGearNoise_Backgrounder.html*.

France

CALIPSO
Cloud-Aerosol LIDAR and Infrared Pathfinder Satellite Observations

The Cloud-Aerosol LIDAR and Infrared Pathfinder Satellite Observation (CALIPSO) satellite, a joint NASA–French National Center for Space Studies (CNES) mission, provides new insight into the role that clouds and atmospheric aerosols (i.e., airborne particles) play in regulating Earth's weather, climate, and air quality. CALIPSO combines an active light detection and ranging (LIDAR) instrument with passive infrared and visible imagers to probe the vertical structure and properties of thin clouds and aerosols over the globe. This data helps to improve our understanding of the role that aerosols and clouds play in regulating the Earth's climate—in particular, how aerosols and clouds interact with one another.

An artist's concept of the CALIPSO satellite orbiting Earth.

CALIPSO was launched by NASA in 2006 with the Canadian Space Agency (CSA)–NASA CloudSat satellite. CALIPSO and CloudSat fly in orbital formation as part of the Afternoon Constellation (A-Train) of Earth-orbiting remote sensing satellites that include NASA's Aqua and Aura spacecraft and CNES's Polarization and Anisotropy of Réflectances for Atmospheric Sciences coupled with Observations from a LIDAR (PARASOL).

CALIPSO provides a unique, 3-year, coincident set of global data on aerosol and cloud properties, radiative fluxes, and atmospheric state. The scientific advances enabled by this coincident dataset improve short-term (measured in days) weather forecasts, as well as forecasts on seasonal-to-interannual timescales. CALIPSO augments the capabilities of the civilian operational Earth-observing satellite used by the National Weather Service, offering a unique capability to monitor volcanic plumes and the long-range transport of pollutants that impact air quality and visibility.

Data from CALIPSO helps scientists to improve predictions of the regional impacts of long-term climate change, ensuring a scientific basis for understanding and assessing the impact of climate change.

CNES provided the Proteus platform, satellite engineering and operations, the Imaging Infrared Radiometer (IIR), algorithm development for the IIR, and a data site for the mission. NASA provided the Wide-Field Camera (WFC), the payload telemetry system to downlink the payload data, the payload onboard computer, and the payload structure. NASA launched the CALIPSO satellite, and it also commands and controls the payload. CNES commands and controls the platform.

For more information on CALIPSO, please visit *http://www.nasa.gov/mission_pages/calipso/main/*.

Human Exploration and Operations

France

DECLIC-DSI
DEvice for the study of Critical LIquids and Crystallization–Directional Solidification Insert

DEvice for the study of Critical LIquids and Crystallization (DECLIC) is a miniature laboratory that was installed on the International Space Station (ISS) in 2009. The French National Center for Space Studies (CNES) and NASA collaborated on the development, launch, and operation of DECLIC–Directional Solidification Insert (DSI).

DECLIC is a multi-user facility that has a range of optical equipment to investigate low- and high-temperature critical fluids behavior, chemical reactivity in supercritical water, directional solidification of transparent alloys, and, more generally, to investigate transparent media in the microgravity environment of the ISS. One of the three modules that were developed for DECLIC is the DSI, which was dedicated to the study of solidification to improve the understanding of metallurgical processes. It involved the in situ and real-time observation of the microstructures that form at the liquid-solid interface when transparent materials solidify. This made it possible to create and transfer images from the investigation back to Earth, instead of creating heavy samples for post mortem characterization that would have required transport from orbit back to Earth.

By the end of November 2010, DECLIC-DSI completed four successful solidifications, and the research team captured 7,000 images during the final session to meet all scientific objectives. The study of solidification microstructure formation is important in the design and processing of new material. Experiments such as DECLIC provide a better understanding of the relationship between micro- and macrostructure formation during solidification processes. The experiment ultimately could result in new and better materials for use in manufacturing on Earth.

For more information on DECLIC-DSI, please visit *http://www.nasa.gov/mission_pages/station/research/experiments/DECLIC-DSI.html* and *http://www.cnes.fr/web/CNES-en/6831-declic.php*.

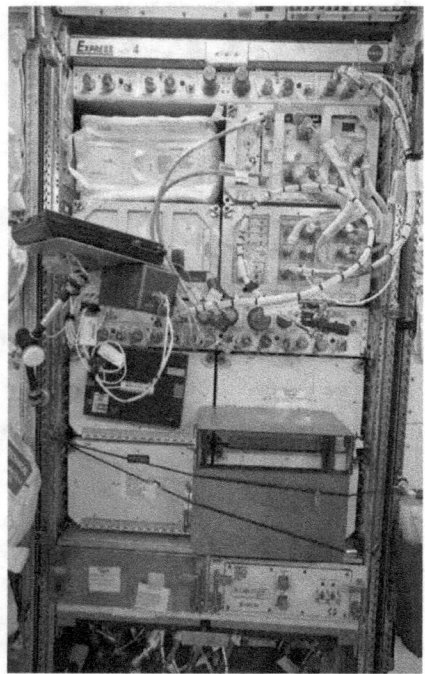

The DECLIC hardware fits into the two lockers in the top right quarter of EXpedite the PRocessing of Experiments to Space Station (EXPRESS) Rack 4 on the International Space Station.

France

Fermi Gamma-ray Space Telescope

The Fermi Gamma-ray Space Telescope, launched by NASA in 2008, is a spacecraft with the ability to detect gamma rays created by the most energetic objects and phenomena in the universe. Among the topics of cosmological interest studied by the mission are dark matter and the periods of star and galaxy formation in the early universe. The mission is an astrophysics and particle physics partnership developed by NASA in collaboration with the U.S. Department of Energy, with important contributions from academic institutions and partners in France, Germany, Italy, Japan, Sweden, and the United States.

Because of their tremendous energy, gamma rays travel through the universe largely unobstructed. This means that Fermi is able to observe gamma-ray sources located near the edge of the visible universe. Gamma rays detected by Fermi originate near the otherwise-obscured central regions of exotic objects like supermassive black holes, pulsars, and gamma-ray bursts.

In order to study these high-energy waves, two main instruments are used: the Large Area Telescope (LAT) and the Gamma-ray Large Area Space Telescope (GLAST) Burst Monitor (GBM). When in survey mode, the LAT scans the entire sky every 3 hours. The instrument detects photons with energies ranging from 20 million electron volts to over 300 billion electron volts. In its first month of operation, the GBM instrument spotted 31 gamma-ray bursts, which occur when massive stars die or when orbiting neutron stars spiral together and merge. By the end of 2011, Fermi had detected more than 100 gamma-emitting pulsars; before its launch, only 7 of these were known to emit gamma rays. Newly developed methods of analysis promise to spur additional discoveries.

An artist's concept of the Fermi spacecraft orbiting Earth.

NASA is working with a number of French institutions on Fermi's LAT Calorimeter, including the Nuclear Studies Center of Bordeaux-Gradignan; the Alternative Energies and Atomic Energy Commission (CEA) Department of Astrophysics, Particle Physics, Nuclear Physics, and Associated Instrumentation; CEA Saclay; the National Institute of Nuclear and Particle Physics (IN2P3); the Laboratory of Theoretical Physics and Astrophysics, Montpellier; and the Laboratoire Leprince-Ringuet, École Polytechnique.

For the Fermi mission, France provided support for the development of the LAT, specifically on the detector that will calculate the amount of energy in the gamma rays. Along with France and NASA, the following organizations are also contributors to the Fermi mission: the German Aerospace Center (DLR), Italian Space Agency (ASI), Japan Aerospace Exploration Agency (JAXA), and the Swedish Royal Institute of Technology (KTH).

For more information on the FERMI mission, please visit *http://fermi.gsfc.nasa.gov/*.

Science

France

InSight
Interior Exploration using Seismic Investigations, Geodesy and Heat Transport

NASA's Interior Exploration using Seismic Investigations, Geodesy and Heat Transport (InSight) mission is designed to improve our understanding of the evolutionary formation of rocky planets, including Earth, by investigating the interior structure and processes of Mars. The mission will also enable scientists to investigate the dynamics of Martian tectonic activity and meteorite impacts, which could offer clues about such phenomena on Earth.

The InSight mission, planned for launch in March 2016 and arrival at Mars in September 2016, consists of a single geophysical lander on Mars to study the planet's deep interior. The InSight science payload comprises two instruments: the Seismic Experiment for Interior Structure (SEIS) and the Heat Flow and Physical Properties Package (HP3). Additionally, the Rotation and Interior Structure Experiment (RISE) will use the spacecraft's communication system to provide precise measurements of planetary rotation. The InSight payload comprises two additional subsystems: the Instrument Deployment System (IDS), which is a robotic arm system for deploying the instruments on the ground; and an Auxiliary Payload Sensor Subsystem (APSS) of wind, temperature, pressure, and magnetic-field sensors.

The French National Center for Space Studies (CNES) is expected to lead an international consortium, including Germany, Spain, Switzerland, and the United Kingdom, to provide the SEIS instrument. SEIS will measure seismic waves traveling through the interior of Mars to determine its interior structure and composition.

The German Aerospace Center (DLR) is expected to lead the development of the HP3 instrument. HP3 will determine the geothermal heat flux of Mars by obtaining thermal measurements in the shallow subsurface, which uniquely constrains the planetary heat engine that drives geologic processes.

For more information on the InSight mission, please visit *http://insight.jpl.nasa.gov/home.cfm*.

An artist's concept of InSight investigating the interior of Mars. InSight is based on the proven NASA Phoenix Mars spacecraft and lander design, with state-of-the-art avionics from the NASA Mars Reconnaissance Orbiter and Gravity Recovery and Interior Laboratory missions. (Credit: NASA/Jet Propulsion Laboratory–Caltech)

France

Juno

NASA's Juno spacecraft, launched in 2011 and slated to arrive in orbit around Jupiter in 2016, will gather data on Jupiter's interior structure, atmospheric composition and dynamics, and polar magnetosphere. This data will be used by scientists to study the origin and evolution of Jupiter, thereby furthering our understanding of planetary and solar-system formation.

Using a spinning, solar-powered spacecraft, Juno will make maps of the gravitational and magnetic fields, auroral emissions, and atmospheric composition of Jupiter from a polar orbit. Juno will carry precise high-sensitivity radiometers, magnetometers, and gravity science systems. Juno's 32 orbits over 11 days will sample Jupiter's full range of latitudes and longitudes. From its polar perspective, Juno will combine in situ and remote sensing observations to explore the polar magnetosphere and determine what drives Jupiter's remarkable auroras. Jupiter's solid core and abundance of heavy metals in the atmosphere make it an ideal model to help improve our understanding of the origin of giant planets. Juno will measure global abundances of oxygen and nitrogen by mapping the gravitational field and using microwave observations of water and ammonia.

The French National Center for Space Studies (CNES) contributed electro-optics for the Juno mission's Jovian Auroral Distributions Experiment (JADE) electron sensor instrument. The JADE instrument will work with some of Juno's other instruments to identify the particles and processes that produce Jupiter's stunning auroras.

For more information on the Juno mission, please visit *http://juno.nasa.gov* and *http://missionjuno.swri.edu/*.

This artist's concept depicts NASA's Juno spacecraft above Jupiter's north pole.

Science

France

MMS
Magnetospheric MultiScale mission

Violent reconnection events can occur as looping lines of magnetic force collide on the surface of the Sun.

NASA's Magnetospheric MultiScale (MMS) mission is a solar-terrestrial probe mission comprising four identically instrumented spacecraft that will use Earth's magnetosphere as a laboratory to study the microphysics of three fundamental plasma processes: magnetic reconnection, energetic particle acceleration, and turbulence. These processes occur in all astrophysical plasma systems but can be studied in situ only in our solar system and most efficiently only in Earth's magnetosphere, where they control the dynamics of the geospace environment and play an important role in the processes known as "space weather." The four MMS spacecraft are currently planned for launch by NASA in 2014.

NASA is collaborating on the MMS mission with the French National Center for Space Studies (CNES), with support from the French Laboratory of Plasma Physics (LPP). NASA is also cooperating with the Aeronautics and Space Agency (ALR) of the Austrian Research Promotion Agency (FFG), the Japan Aerospace Exploration Agency (JAXA), and the Swedish National Space Board (SNSB) on the mission.

MMS investigates how the magnetic fields of the Sun and Earth connect and disconnect, explosively transferring energy from one to the other in a process known as magnetic reconnection. This process limits the performance of fusion reactors and is the final governor of geospace weather, which affects modern technological systems such as telecommunications networks, Global Positioning System (GPS) navigation, and electrical power grids. MMS's four spacecraft will measure plasmas, fields, and particles in a near-equatorial orbit that will frequently encounter reconnection in action.

The four MMS spacecraft will carry identical instrument suites of plasma analyzers, energetic particle detectors, magnetometers, and electric field instruments; they will also carry a device to prevent spacecraft charging from interfering with the highly sensitive measurements required in and around the diffusion regions. CNES is designing and developing the Search Coil Magnetometers (SCM) and support electronics for the four MMS spacecraft. CNES will also support the integration and testing of the SCM. NASA is providing the spacecraft, the launch, and overall mission management.

For more information on MMS, please visit *http://mms.gsfc.nasa.gov/*.

France

MAVEN
Mars Atmosphere and Volatile Evolution Mission

NASA's Mars Atmosphere and Volatile Evolution (MAVEN) mission, launched in November 2013, will orbit Mars for 1 year and provide the first direct measurements to address key scientific questions about Mars's evolution. There is evidence that Mars once had a denser atmosphere that supported the presence of liquid water on the surface. As part of a dramatic climate change, Mars lost most of its atmosphere. MAVEN is expected to make definitive scientific measurements of processes controlling Mars's present-day atmospheric loss that will offer clues about the planet's atmospheric history.

The French National Center for Space Studies (CNES) provided the Solar Wind Electron Analyzer (SWEA), a component of the Particles and Fields Package, to measure solar wind and ionospheric electrons. CNES provided similar SWEA analyzers on the Solar TErrestrial RElations Observatory (STEREO) mission.

MAVEN carries a payload of eight instruments: the SWEA, the Solar Wind Ion Analyzer (SWIA), the SupraThermal and Thermal Ion Composition (STATIC) instrument, the Solar Energetic Particle detector, the Langmuir Probe and Waves (LPW) instrument, a magnetometer, an Imaging Ultraviolet Spectrograph (IUVS), and a Neutral Gas and Ion Mass Spectrometer (NGIMS).

For more information on the MAVEN mission, please visit *http://mars.jpl.nasa.gov/programmissions/missions/future/maven/* and *http://lasp.colorado.edu/home/maven/*.

Engineers inspect the MAVEN primary structure, which is built out of composite panels of aluminum honeycomb sandwiched between graphite composite face sheets and attached to one another with metal fittings. The entire structure weighs only 275 pounds (125 kilograms). (Credit: NASA/Lockheed Martin)

Science

France

Mars Odyssey

The French National Center for Space Studies (CNES), as a member of the Gamma Ray Spectrometer (GRS) Science Team, provided four germanium detectors for the GRS instrument suite and science support for NASA's Mars Odyssey mission, which was launched in 2001.

The Mars Odyssey spacecraft mapped, for the first time, the amount and distribution of chemical elements and minerals that make up the Martian surface. Odyssey's maps of hydrogen distribution led scientists to discover vast amounts of water ice in the polar regions of Mars, buried just beneath the surface. Odyssey also recorded the radiation environment in low-Mars orbit to determine the radiation-related risk to future human explorers who may one day go to Mars.

The GRS instrument suite—one of three primary instrument packages on the Mars Odyssey—has collected data on the distribution of elements at the surface of Mars and searched for water across the planet. The GRS instrument suite detects and counts gamma rays and neutrons from the Martian surface. Scientists use the GRS data to calculate the ratio of elemental abundances on the surface by associating the spectral distribution of the gamma rays with known nuclear transitions lines. By counting the number of neutrons as a function of energy, it is possible to calculate the hydrogen abundance, thus inferring the presence of water.

Odyssey is one of five active robotic missions at Mars, which include NASA's Mars Reconnaissance Orbiter (MRO), Mars Exploration Rover (MER) (Opportunity), the Mars Science Laboratory (MSL), and the European Space Agency's (ESA) Mars Express orbiter. The orbiters and surface missions complement each other in many ways. Observations by the rovers provide ground-level understanding for interpreting global observations by the orbiters. In addition to their own science missions, the orbiters relay data from the Mars Rovers.

For more information on Mars Odyssey, please visit *http://marsprogram.jpl.nasa.gov/odyssey/overview/*.

This artist's rendering (not to scale) portrays ice-rich layers in the soils of Mars being detected by the Gamma-Ray Spectrometer suite of instruments aboard NASA's 2001 Mars Odyssey spacecraft. Measurements indicate that the upper meter (3 feet) of soil in the polar regions contains an ice-rich zone with an ice abundance of 20 to 50 percent by mass. (Credit: NASA/JPL/University of Arizona/Los Alamos National Laboratories)

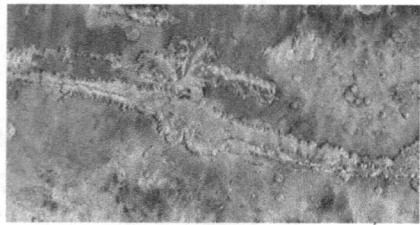

This mosaic image of the Valles Marineris canyon, colored to resemble the Martian surface, comes from the Thermal Emission Imaging System (THEMIS), a visible-light and infrared-sensing camera on the Mars Odyssey orbiter.

France

MSL
Mars Science Laboratory

NASA's Mars Science Laboratory (MSL) mission utilized an innovative sky crane to set down the Curiosity rover onto the surface of Mars in August 2012, where Curiosity began its mission to assess whether Mars has ever had an environment capable of supporting small life-forms called microbes. The MSL mission, launched by NASA in November 2011, is planned to last at least one Martian year (687 Earth days), but it has the potential to operate for much longer.

This image shows the increasing wheel size among the Mars explorers: Mars Pathfinder (left), Mars Exploration Rover (center), and Mars Science Laboratory (right).

The primary scientific objective, to be carried out during the surface science phase of the MSL mission, is to assess the biological potential of at least one target area by characterizing the local geology and geochemistry, investigating planetary processes relevant to habitability (including the role of water), and characterizing the broad spectrum of surface radiation. Since its landing, MSL has made new and exciting science discoveries, such as observing a natural intersection of three kinds of terrain at an area called Glenelg and finding evidence that a stream once ran across the area on Mars where the rover is driving. Although earlier missions detected the presence of water on Mars, the images taken by MSL of rocks that contain ancient streambed gravel are the first of their kind.

The French National Center for Space Studies (CNES) provided significant hardware for two of MSL's instrument suites: the Laser-Induced Remote Sensing for Chemistry and Micro-Imaging (ChemCam) and the Sample Analysis at Mars (SAM). For the ChemCam instrument suite, CNES provided the ChemCam mast unit, including the laser, telescope, camera, and front-end electronics. The ChemCam instrument suite is used to characterize the geology of the Mars landing region, investigate planetary processes relevant to past habitability, assess the biological potential of a target environment, and look for toxic materials. Looking at rocks and soils from a distance, ChemCam fires its laser to analyze the elemental composition of vaporized materials from areas smaller than 1 millimeter on the surface of Martian rocks and soils. An onboard spectrograph provides detail about minerals and microstructures in rocks by measuring the composition of the resulting plasma—a visible brief glow of ionized material.

CNES provided the gas chromatograph portion of the SAM instrument suite. SAM is used to study geochemical conditions that are directly relevant to the larger goal of assessing the habitability of Mars. SAM also searches for compounds of the element carbon, including methane, that are associated with life and explores ways in which they are generated and destroyed in the Martian ecosphere.

For more information on the MSL mission and the ChemCam and SAM instrument suites, please visit *http://mars.jpl.nasa.gov/msl/*, *http://marsprogram.jpl.nasa.gov/msl/mission/instruments/spectrometers/chemcam/*, and *http://marsprogram.jpl.nasa.gov/msl/mission/instruments/spectrometers/sam/*.

Science

An artist's rendering of the OSTM satellite orbiting Earth. (Credit: NASA/Jet Propulsion Laboratory–Caltech)

France

OSTM
Ocean Surface Topography Mission

The Ocean Surface Topography Mission (OSTM), launched by NASA in 2008, provides continuity of ocean topography measurements beyond the 1992 Topex/POSEIDON and 2001 Jason-1 missions for determining ocean circulation, climate change, sea-level rise, and societal applications (e.g., El Niño and hurricane forecasting).

OSTM functions as a bridge to an operational mission for the continuation of multidecadal ocean topography measurements. OSTM is a joint project between the French National Center for Space Studies (CNES), NASA, the U.S. National Oceanic and Atmospheric Administration (NOAA), and the European Organisation for the Exploitation of Meteorological Satellites (EUMETSAT). This mission illustrates a clear transition from research and development agencies to operational agencies for ocean surface topography measurements in both Europe and the United States. CNES provided the satellite platform, payload module, a Poseidon-3 dual-frequency radar altimeter, the Doppler Orbitography and Radiopositioning Integrated by Satellite (DORIS) receiver package, and DORIS auxiliary instruments. NASA provided the microwave radiometer, laser retro-reflector, and Global Positioning System (GPS) receiver package, and it also launched the satellite.

NOAA established an operations control center, along with communication and data acquisition stations, and it operates and controls the OSTM satellite. NOAA also provides operational data processing, distribution, and archiving. EUMETSAT provides an Earth terminal, operational data processing, and distribution to European users.

For more information on OSTM, please visit *http://sealevel.jpl.nasa.gov/* and *http://www.aviso.oceanobs.com/*.

France

SPP
Solar Probe Plus

NASA's Solar Probe Plus (SPP) mission is a project in the Living With a Star (LWS) program, a series of missions designed to gather critical information about the Sun and its effects on Earth, human activities, and other planetary systems. By flying into the Sun's outer atmosphere, SPP will gather data on the processes that heat the corona and accelerate the solar wind, solving two fundamental mysteries that have been top-priority science goals for many decades. The French National Center for Space Studies (CNES) is collaborating with NASA on the Solar Wind Electrons Alphas and Protons (SWEAP) instrument and the SPP Fields experiment. Belgium and Germany are also participating on this mission.

An artist's concept of the Solar Probe Plus spacecraft, fully deployed in cruise configuration and flying past Venus during one of the seven gravity assists that will send it closer to the Sun. (Credit: Johns Hopkins University Applied Physics Laboratory)

The SPP mission, scheduled to launch in 2018, is planned to last almost 7 years, and the spacecraft will eventually come within 3.7 million miles from the surface of the Sun, well within the orbit of Mercury and about eight times closer than any spacecraft has ever approached. The SPP's orbit will provide it with a unique vantage point to study the streams of charged particles that the Sun hurls into space from the region where the processes that heat the corona and produce solar wind actually occur. At its closest approach SPP will zip past the Sun at 125 miles per second, protected by a carbon-composite heat shield that must withstand up to 2600° Fahrenheit and survive blasts of radiation and energized dust at levels not experienced by any previous spacecraft. The primary scientific objectives to be carried out during the SPP mission include: determining the structure and dynamics of the magnetic fields at the sources of both fast and slow solar wind, tracing the flow of energy that heats the corona and accelerates the solar wind, and determining what mechanisms accelerate and transport energetic particles. Instruments include a wide-field imager, fast-ion analyzer, fast-electron analyzer, energetic-particle instrument, magnetometer, and plasma-wave instrument.

CNES contributions to the SPP mission include the design, fabrication, and delivery of the Thermal Noise Receiver (TNR)/High-Frequency Receiver electronics and the Search Coil Magnetometer (SCM) sensor and preamplifier for the Fields experiment, as well as testing and evaluation of SWEAP instrument components. The Fields experiment consists of two instruments: a plasma-wave instrument and a magnetometer. The Fields experiment will measure the solar wind plasma density using several complementary techniques and at a very high time resolution. These instruments will be used to study how the Sun's corona is heated and accelerated to produce the solar wind. The SWEAP investigation consists of two instruments: a solar probe cup and a solar probe analyzer. The SWEAP instruments will count the most abundant particles in the solar wind (electrons, protons, and helium ions), measure their properties, and capture some of the particles for direct analysis.

For more information on SPP, please visit *http://solarprobe.gsfc.nasa.gov/*.

Science

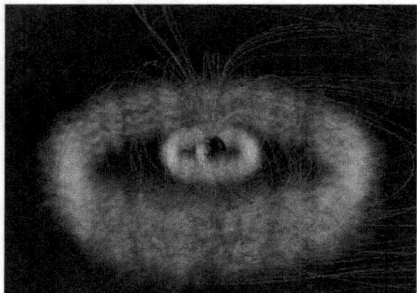

Two giant rings of charged particles called the Van Allen radiation belts surround Earth. The belts pose a hazard to satellites, which must protect their sensitive components with adequate shielding if their orbit requires significant time in the radiation belts. The Space Environment Testbeds will improve hardware performance in the space-radiation environment.

France

SET-1
Space Environment Testbed-1

The Space Environment Testbed-1 (SET-1) project performs flight and ground investigations to address the Living With a Star (LWS) program's goal of understanding how interactions between the Sun and Earth affect humanity. The SET-1 project is the element of the LWS program that characterizes the space environment and its impact on hardware performance in space. The goal of the SET-1 project is to improve the engineering approach to accommodation and/or mitigation of the effects of solar variability on spacecraft design and operations. The French National Center for Space Studies (CNES) is collaborating with NASA on this project. International participation also includes the United Kingdom Space Agency.

The primary science objectives for the SET-1 project are to define space environment effects and mechanisms, to reduce design margins, and to improve design and operations guidelines. When margins for spacecraft design and operations are reduced, "space-environment-tolerant" and new technologies can be used more frequently, the fraction of spacecraft resources available for payloads can be increased or launch-vehicle requirements can be reduced, and routine operations above low-Earth orbit (LEO) can be performed at the same cost as LEO operations.

NASA and CNES are conducting the Commercial Off-The-Shelf-2 (COTS-2) digital technologies experiment as part of the SET-1 project. The experiment will be used to measure the effects of the space ionizing radiation environment on COTS-2 digital microelectronics devices in order to improve performance prediction in space for future digital devices. CNES will be providing for the design and the breadboard fabrication of the COTS-2 digital microelectronics devices experiment. The SET-1 will be flown on the Air Force Research Laboratory Demonstration and Science Experiments (DSX) spacecraft and is scheduled to launch in 2014.

For more information on SET-1, please visit *http://science.nasa.gov/missions/space-environment-testbeds/*.

France

SWOT
Surface Water Ocean Topography

The joint NASA–French National Center for Space Studies (CNES) Surface Water Ocean Topography (SWOT) mission, planned for launch in 2020, will improve our understanding of the world's oceans and its terrestrial surface waters. It will use wide-swath altimetry to measure spatial fields of surface elevations for both inland waters and the oceans. This will lead to new information about the dynamics of water stored at the land's surface (in lakes, reservoirs, wetlands, and river channels) and improved estimates of deep-ocean and near-coastal marine circulation.

Plans are for NASA to provide the payload module, the Ka-band Radar Interferometer (KaRIn) Microwave Radiometer (MR) with its antenna, a laser retroreflector array (LRA), a Global Positioning System (GPS) receiver package, and all launch services. NASA will also provide a payload control center and ground terminals, as well as data processing, archiving, and distribution infrastructure for the mission. CNES plans to provide the spacecraft bus, the dual-frequency (Ku/C band) Nadir Altimeter, the Doppler Orbitography and Radiopositioning Integrated by Satellite (DORIS) receiver package, and the Radio Frequency Unit (RFU) for the KaRIn instrument. CNES may also provide a command and control center for the satellite, ground terminals, and data processing, archiving, and distribution infrastructure for the mission. Through NASA, the Canadian Space Agency (CSA) plans to provide Extended Interaction Klystrons (EIK) for the KaRIn.

For more information on SWOT, please visit *http://eospso.gsfc.nasa.gov/missions/surface-water-ocean-topography* and *http://swot.jpl.nasa.gov/*.

An artist's depiction of SWOT orbiting Earth.

Science

The highly distorted supernova remnant shown in this image may contain the most recent black hole formed in the Milky Way galaxy. The image combines x rays from NASA's Chandra X-ray Observatory (blue and green), radio data from the National Science Foundation's Very Large Array (pink), and infrared data from Caltech's Palomar Observatory (yellow).

Chandra observations of the massive spiral galaxy NGC 5746 revealed a large halo of hot gas (blue) surrounding the optical disk of the galaxy (white). The halo extends more than 60,000 light-years on either side of the disk of the galaxy.

Germany

Chandra X-ray Observatory

The Chandra X-ray Observatory, launched on the Space Shuttle in 1999, is part of NASA's fleet of "Great Observatories" along with the Hubble Space Telescope and the Spitzer Space Telescope. Chandra allows scientists from around the world to obtain unprecedented x-ray images of exotic environments in order to help understand the structure and evolution of the universe. Having already surpassed its originally planned 5-year lifetime, Chandra's observations are rewriting textbooks and helping to advance technology.

The Chandra telescope system consists of four pairs of mirrors and their support structure. X-ray telescopes function differently than optical telescopes in that high-energy x-ray photons penetrate into a mirror much as bullets slam into a wall, and then the photons ricochet off the mirrors. The shape of the mirrors directs the photons to one spot, where the four scientific instruments then analyze information from these photons.

The German Aerospace Center (DLR) is cooperating with NASA on Chandra by providing contributions to the Electron, Proton, Helium Instrument (EPHIN) and the Low-Energy Transmission Grating for Cosmic X-ray Spectrometer (LETGS). The EPHIN detector monitors the local charged particle environment as part of the scheme to protect the two focal-plane science instruments from particle radiation. DLR also sponsored the participation of the German scientist for the EPHIN cooperation.

For more information on Chandra, please visit *http://www.nasa.gov/mission_pages/chandra/main/* and *http://chandra.harvard.edu/*.

Germany

CFD
Computational Fluid Dynamics

NASA and the German Aerospace Center (DLR) began cooperating on enhancing computational fluid dynamics (CFD) capabilities for flow and acoustic field simulations and validation studies in 2007. This cooperation enables NASA and DLR to engage in advanced modeling and simulation research that will be critical for the development of more predictive and accurate fluid dynamics simulations.

NASA-DLR cooperation on CFD activities include: research that contributes to the development of improved prediction methods using high-fidelity simulations of unsteady flow fields, and to the development of validated physics-based multidisciplinary design and analysis tools for aircraft; projects that contribute to the development of validated simulations and techniques for high-fidelity analysis of fluid flow phenomena and that focus on advanced flow simulation capabilities for higher-order methods, large eddy simulations, and hybrid methods for nonequilibrium flows; and research that contributes to noise reduction, environmental research using aero-acoustic simulation capabilities and methods, and validation of employed approaches to model generation of turbulence and noise.

For more information on NASA's computational fluid dynamics research, please visit *http://www.aeronautics.nasa.gov/fap/aboutus.html*.

A CFD validation test of the DLR F6 Commercial Aircraft Configuration in the National Transonic Facility at NASA Langley Research Center.

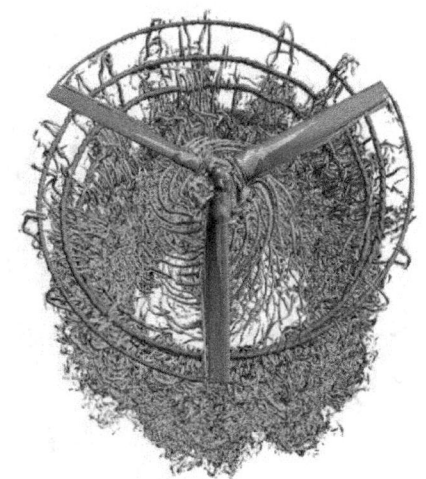

A Navier-Stokes simulation of a V-22 Osprey rotor in hover; magenta indicates high vorticity, and blue indicates low vorticity.

Science

Germany

Dawn Asteroid Rendezvous Mission

NASA's Dawn mission, launched in 2007, sent an orbiting space probe to examine the dwarf planet Ceres and the asteroid Vesta, two of the most massive members of the asteroid belt. In addition to being the first NASA science mission to make use of ion propulsion, Dawn will also be the first mission to visit and orbit two distinct bodies other than Earth and the Moon.

Dawn's goal is to characterize the conditions and processes of the solar system's earliest epoch by investigating in detail two of the largest protoplanets that have remained relatively intact since their formations. Together with many other smaller bodies, Ceres and Vesta reside in the extensive zone between Mars and Jupiter called the asteroid belt. Each has followed a very different evolutionary path, constrained by the diversity of processes that operated during the first few million years of solar system evolution.

The Dawn spacecraft successfully entered into orbit around Vesta in July 2011, and it has since revealed that the asteroid is a layered planetary building block with an iron core—the only one known to have survived the earliest days of the solar system. Dawn departed Vesta in September 2012 and will begin to study Ceres upon its arrival in 2015.

The German Aerospace Center (DLR) developed, supported, and integrated two framing cameras (FC) that provide scientific imagery for the Dawn mission.

For more information on the Dawn mission, please visit *http://dawn.jpl.nasa.gov/mission/index.asp*.

This artist's rendering shows NASA's Dawn spacecraft orbiting the giant asteroid Vesta. The depiction of Vesta is based on images obtained by Dawn's framing cameras. (Credit: NASA/Jet Propulsion Laboratory–Caltech)

Scientists at the Jet Propulsion Laboratory inspect the Dawn spacecraft.

Germany

Fermi Gamma-ray Space Telescope

The Fermi Gamma-ray Space Telescope, launched by NASA in 2008, is a spacecraft with the ability to detect gamma rays created by the most energetic objects and phenomena in the universe. Among the topics of cosmological interest studied by the mission are dark matter and the periods of star and galaxy formation in the early universe. The mission is an astrophysics and particle physics partnership, developed by NASA in collaboration with the U.S. Department of Energy, with important contributions from academic institutions and partners in France, Germany, Italy, Japan, Sweden, and the United States. NASA is working with the German Aerospace Center (DLR), which is supporting the German Max Planck Institute for Extraterrestrial Physics (MPE) on the Fermi mission.

The Fermi spacecraft shortly before launch. The solar panels are folded at the sides. The GBM detector modules and the telemetry antennas can be seen on the left side.

Because of their tremendous energy, gamma rays travel through the universe largely unobstructed. This means that Fermi is able to observe gamma-ray sources located near the edge of the visible universe. Gamma rays detected by Fermi originate near the otherwise-obscured central regions of exotic objects like supermassive black holes, pulsars, and gamma-ray bursts.

In order to study these high-energy waves, two main instruments are used: the Large Area Telescope (LAT) and the Gamma-ray Large Area Space Telescope (GLAST) Burst Monitor (GBM). When in survey mode, the LAT scans the entire sky every 3 hours. The instrument detects photons with energies ranging from 20 million electron volts to over 300 billion electron volts. In its first month of operation, the GBM instrument spotted 31 gamma-ray bursts, which occur when massive stars die or when orbiting neutron stars spiral together and merge. By the end of 2011, Fermi had detected more than 100 gamma-emitting pulsars; before its launch, only 7 of these were known to emit gamma rays. Newly developed methods of analysis promise to spur additional discoveries.

For the Fermi mission, through DLR's support, the Max Planck Institute built all of the sensors and the power-supply box and support on the GBM instrument. One of its scientists is the GBM co-principal investigator. Along with DLR and NASA, the following organizations are also contributors to the Fermi mission: the French Alternative Energies and Atomic Energy Commission (CEA), Italian Space Agency (ASI), Japan Aerospace Exploration Agency (JAXA), and the Swedish Royal Institute of Technology (KTH).

For more information on the Fermi mission, please visit *http://fermi.gsfc.nasa.gov/*.

Science

An artist's concept of the twin GRACE satellites orbiting Earth.

Germany

GRACE
Gravity Recovery and Climate Experiment

The Gravity Recovery and Climate Experiment (GRACE) is a NASA Earth System Science Pathfinder mission that was initiated through a partnership with the German Aerospace Center (DLR). GRACE consists of a pair of identical satellites in coplanar formation flight whose relative velocities are measured with submicron-per-second velocities over a range of more than 200 kilometers. This tandem formation provides spatial and temporal gravity field measurements of a few micrograms (a few parts per billion) to measure the movement of mass within the Earth system.

In 2003, using only 111 days of GRACE satellite data, the GRACE science team released a preliminary model of Earth's gravity field 10 to 100 times more accurate than the previous model, which had been constructed from decades of geodetic data. GRACE improvements to the gravity field data have advanced numerous fields of endeavor, from the observation of the Lense-Thirring effect of general relativity to improvements in continental scale leveling for mapping and development projects. The gravity field of Earth varies both in space and time according to the movement of mass due to tectonic forces, ocean and atmospheric circulation, and changing hydrology from weather and climate changes. GRACE's time-variable gravity data is yielding crucial information about the distribution and movement of mass within the solid Earth and its fluid surroundings, such as estimated post-glacial rebound rates, changes in the polar ice sheets and mountain glaciers, changes in water distribution within the world's major drainage basins, changes in deep ocean circulation, and, finally, the relative contributions of ocean warming and water mass distribution to sea level change.

NASA provided the two GRACE spacecraft, Microwave Ranging Instrument (MRI), accelerometers, and Global Positioning System (GPS) hardware. DLR provided the launch and, through its German Space Operations Center (GSOC), is providing mission operations, controlling the satellite flight operations, and providing data downlink and archive facilities.

GRACE is helping oceanographers unlock the secrets of ocean circulation and its effect on climate. Launched on March 17, 2002, from Plesetsk Cosmodrome in Arkhangelsk Oblast, Russia, GRACE is currently in an extended period of operations beyond its planned mission life of 5 years.

For more information on GRACE, please visit *http://www.csr.utexas.edu/grace/* and *http://www.dlr.de/*.

Germany

GRACE-FO
Gravity Recovery and Climate Experiment Follow-On

The Gravity Recovery and Climate Experiment Follow-On (GRACE-FO), a partnership between NASA and the German Research Centre for Geosciences (GFZ), will provide continuity of measurements acquired by the 2002 Gravity Recovery and Climate Experiment mission. These measurements provide high-resolution global data on the Earth's gravity field, including how this field varies over time.

GRACE-FO will consist of a pair of identical satellites in coplanar formation flight whose relative velocities are measured with submicron-per-second velocities over a range of more than 200 kilometers. This tandem formation provides spatial and temporal gravity-field measurements of a few micrograms (a few parts per billion) to measure the movement of mass within the Earth system. The mission may also demonstrate a new laser ranging technology to support future gravity missions.

NASA will provide the two GRACE-FO spacecraft, Microwave Ranging Instrument (MRI), accelerometers, and Global Positioning System (GPS) hardware, and it will lead joint development with GFZ of a Laser Ranging Interferometer (LRI). GFZ will provide the launch, mission operations, control of the satellite flight operations, and data downlink and archive facilities. GRACE-FO will be launched in the 2017 timeframe.

For more information on GRACE-FO, please visit *http://www.nasa.gov/grace/* and *http://grace.jpl.nasa.gov/*.

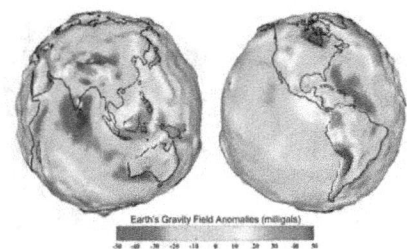

Due to an uneven distribution of mass inside the planet, Earth's gravity field is not uniform—that is, it has "lumps." This model greatly exaggerates the scale so that smaller features can be seen. The GRACE Follow-On mission will continue the GRACE mission's mapping of the precise location and size of these lumps, enabling a greater understanding of the structure of Earth. (Credit: NASA/University of Texas Center for Space Research)

Science

An artist's rendering of the formation of rocky bodies in the solar system—how they form, differentiate, and evolve into terrestrial planets. (Credit: NASA/Jet Propulsion Laboratory–Caltech)

Germany

InSight
Interior Exploration using Seismic Investigations, Geodesy and Heat Transport

NASA's Interior Exploration using Seismic Investigations, Geodesy and Heat Transport (InSight) mission is designed to improve our understanding of the evolutionary formation of rocky planets, including Earth, by investigating the interior structure and processes of Mars. The mission will also enable scientists to investigate the dynamics of Martian tectonic activity and meteorite impacts, which could offer clues about such phenomena on Earth.

The InSight mission, planned for launch in March 2016 and arrival at Mars in September 2016, consists of a single geophysical lander on Mars to study the planet's deep interior. The InSight science payload comprises two instruments: the Seismic Experiment for Interior Structure (SEIS) and the Heat Flow and Physical Properties Package (HP3). Additionally, the Rotation and Interior Structure Experiment (RISE) will use the spacecraft's communication system to provide precise measurements of planetary rotation. The InSight payload includes two additional subsystems: the Instrument Deployment System (IDS), which is a robotic arm system for deploying the instruments on the ground; and an Auxiliary Payload Sensor Subsystem (APSS) of wind, temperature, pressure, and magnetic-field sensors.

The German Aerospace Center (DLR) is expected to lead the development of the HP3 instrument. HP3 will determine the geothermal heat flux of Mars by obtaining thermal measurements in the shallow subsurface that uniquely constrain the planetary heat engine that drives geologic processes.

The French National Center for Space Studies (CNES) is expected to lead an international consortium, including Germany, Spain, Switzerland, and the United Kingdom, to provide the SEIS instrument. SEIS will measure seismic waves traveling through the interior of Mars to determine its interior structure and composition.

For more information on the InSight mission, please visit *http://insight.jpl.nasa.gov/home.cfm*.

Germany

MER
Mars Exploration Rovers

The Mars Exploration Rover (MER) mission, involving two robotic geologists—Spirit and Opportunity—is part of NASA's Mars Exploration Program, a long-term robotic exploration of the Red Planet. The rovers were launched separately in 2003 and landed in different areas of Mars in January 2004.

Primary among the MER mission's scientific goals is to search for and characterize a wide range of rocks and soils that hold clues to past water activity. The MER mission seeks to determine where conditions may once have been favorable to life by analyzing the climate and water histories at sites on Mars. Each rover is equipped with the Athena science payload, which is used to read the geological record at each site, investigate what role water played there, and determine how suitable the conditions would have been for life.

An artist's concept of an MER on the Martian surface.

The German Aerospace Center (DLR) provided two science instruments for the Athena payload: the Alpha Particle X-ray Spectrometer (APXS), which determines the elemental composition of rocks and soils while providing scientists with information about crustal formation and weathering processes; and the Mössbauer Spectrometer (MB), which determines the composition and abundance of iron-bearing minerals and examines the magnetic properties of surface materials. The MB also provides information about early Martian environmental conditions and possible fossil evidence.

DLR also sponsored three German scientists for the MER Science Team, who are using MER data to research soil composition at both global and local scales, investigate Martian dust, and research Martian soil mechanics, including soil perturbations and soil characteristics.

Spirit and Opportunity worked on Mars for well beyond their originally planned 3-month missions: Spirit fell silent in March 2010, but Opportunity has continued to send data downlinks to Earth. Within 2 months of landing on Mars in early 2004, Opportunity found geological evidence of an environment that was once wet. Years of analysis of observations made in 2005 by Spirit's APXS and MB produced a 2011 report that an outcrop on Husband Hill bears a high concentration of carbonate, providing evidence of a wet, nonacidic ancient environment that may have been favorable for microbial life.

For more information on MER, please visit *http://marsrovers.nasa.gov/home/index.html* and *http://athena.cornell.edu/the_mission*.

Science

This artist's concept depicts the moment immediately after NASA's Curiosity rover, lowered via a unique sky-crane system, touched down onto the Martian surface. After the spacecraft detected the touchdown, pyrotechnic cutters severed the connections between the rover and the spacecraft's descent stage, which then flew away, coming back to the surface a safe distance away. (Credit: NASA/Jet Propulsion Laboratory–Caltech)

Germany

MSL
Mars Science Laboratory

NASA's Mars Science Laboratory (MSL) mission utilized an innovative sky crane to set down the Curiosity rover onto the surface of Mars in August 2012, where it began its mission to assess whether Mars has ever had an environment capable of supporting small life forms called microbes. The MSL mission, launched by NASA in November 2011, is planned to last at least one Martian year (687 Earth days), but it has the potential to operate for much longer.

The primary scientific objective, to be carried out during the surface science phase of the MSL mission, is to assess the biological potential of at least one target area by characterizing the local geology and geochemistry, investigating planetary processes relevant to habitability (including the role of water), and characterizing the broad spectrum of surface radiation. The landing site, Gale Crater, was selected based on an assessment of safety and planetary protection and an analysis by the scientific community. Since its landing, MSL has already begun to make new and exciting science discoveries, such as observing a natural intersection of three kinds of terrain at an area called Glenelg and finding evidence that a stream once ran across the area on Mars where the rover is driving. Although earlier missions detected the presence of water on Mars, the images taken by MSL of rocks that contain ancient streambed gravel are the first of their kind.

The German Aerospace Center (DLR) provided significant input to the Radiation Assessment Detector (RAD), which characterizes the broad spectrum of radiation at the surface of Mars—an essential precursor to human exploration of the planet. The RAD also measures and identifies all high-energy radiation on the Martian surface, such as protons, energetic ions of various elements, neutrons, and gamma rays. This includes not only direct radiation from space but also secondary radiation produced by the interaction of space radiation with the Martian atmosphere and surface rocks and soils.

For more information on the MSL mission and the RAD instrument, please visit *http://mars.jpl.nasa.gov/msl/* and *http://marsprogram.jpl.nasa.gov/msl/mission/instruments/radiationdetectors/rad/*.

Germany

SPP
Solar Probe Plus

NASA's Solar Probe Plus (SPP) mission is a project in the Living With a Star (LWS) program, a series of missions designed to gather critical information about the Sun and its effects on Earth, human activities, and other planetary systems. By flying into the Sun's outer atmosphere, SPP will gather data on the processes that heat the corona and accelerate the solar wind, solving two fundamental mysteries that have been top-priority science goals for many decades. The German Aerospace Center (DLR) is collaborating with NASA on the Wide-field Imager for Solar Probe (WISPR) investigation on SPP. Belgium and France are also participating on this mission.

NASA plans to launch SPP in 2018 from Cape Canaveral Air Force Station (CCAFS) in Florida. The mission is planned to last almost 7 years and the spacecraft will eventually come within 3.7 million miles from the surface of the Sun, well within the orbit of Mercury and about eight times closer than any spacecraft has ever approached.

SPP will study the streams of charged particles that the Sun hurls into space from a vantage point where the processes that heat the corona and produce solar wind actually occur. At its closest approach SPP will zip past the Sun at 125 miles per second, protected by a carbon-composite heat shield that must withstand up to 2600° Fahrenheit and survive blasts of radiation and energized dust at levels not experienced by any previous spacecraft. The primary scientific objectives to be carried out during the SPP mission include: determining the structure and dynamics of the magnetic fields at the sources of both fast and slow solar wind, tracing the flow of energy that heats the corona and accelerates the solar wind, and determining what mechanisms accelerate and transport energetic particles. Instruments include a wide-field imager, fast-ion analyzer, fast-electron analyzer, energetic-particle instrument, magnetometer, and plasma-wave instrument.

DLR contributions to the SPP mission include analysis and evaluation of the WISPR investigation and the development of software for WISPR data analysis and modeling tasks. The WISPR investigation is a single visible-light telescope with a wide field of view that will track density fluctuations in the solar corona by imaging visible sunlight scattered by electrons in the corona as the spacecraft traverses through its perihelion passes. Fluctuations can arise from dynamic events, such as coronal mass ejections, and also from the "quiescent" slow and fast solar wind. In addition, the rapid motion of the spacecraft through the corona will result in significant apparent changes due to shifts in its viewpoint, enabling tomographic reconstruction of the structures.

For more information on SPP, please visit *http://solarprobe.gsfc.nasa.gov/*.

An artist's depiction of Solar Probe Plus as it deploys its solar arrays and begins drawing power for its trip toward the Sun. (Credit: Johns Hopkins University Applied Physics Laboratory)

Science

Germany

SOFIA
Stratospheric Observatory for Infrared Astronomy

NASA and the German Aerospace Center (DLR) are collaborating on the Stratospheric Observatory for Infrared Astronomy (SOFIA). SOFIA is not a spacecraft but rather consists of a 2.5-meter telescope aboard a modified Boeing 747SP. Not only can it study the universe in the infrared and submillimeter spectrum, but SOFIA is also integral to the development of new observation techniques and instrumentations.

As a flying observatory, SOFIA consists of two main sections: the modified Boeing 747SP and the telescope. The telescope accommodates a wide variety of scientific instruments operating across a wide range of wavelengths. One of only 45 747SPs ever produced, the SOFIA aircraft was used for commercial passenger transport through 1995. The "SP" designation means "special performance," a reference to the extended range of the aircraft that was achieved via a shorter body than other 747 models. NASA acquired the SOFIA aircraft in 1997 and began large modifications in 1998 to transform it into a flying observatory.

The telescope employs three main mirrors, the largest a 2.7-meter concave mirror that initially captures the light. After reflecting off the other two mirrors, an image is focused on the focal point, where the instruments capture and analyze the data. SOFIA's science instruments include cameras, spectrometers, and photometers that together operate across a very wide range of infrared wavelengths. Once the information is gathered, scientists can use it to analyze black holes, the birth and death of stars, the formation of solar systems, and various other space phenomena.

DLR is contributing the telescope, along with two of the six first-generation instruments. DLR also provides additional operation support during the course of the resulting flights.

For more information on SOFIA, please visit *http://www.sofia.usra.edu/index.html*.

With the large door over its 2.5-meter, German-built telescope wide open, NASA's Stratospheric Observatory for Infrared Astronomy 747SP aircraft soars over southern California's high desert.

A close-up of the SOFIA telescope in flight.

Germany

THEMIS
Time History of Events and Macroscale Interactions during Substorms
and
ARTEMIS
Acceleration, Reconnection, Turbulence and Electrodynamics of the Moon's Interaction with the Sun

An artist's concept of ARTEMIS P1 and P2 orbiting the Moon. (Credit: NASA/Conceptual Image Lab)

NASA's Time History of Events and Macroscale Interactions during Substorms (THEMIS) mission originated as a 2-year mission consisting of five identical satellites, or probes, launched together in 2007 to study the violent and colorful eruptions in auroras. The mission incorporates a network of ground-based auroral observatories.

THEMIS helps to determine which physical processes in near-Earth space initiate violent substorm eruptions in the Earth's magnetosphere. Substorms intensify auroras and create a dramatic "dancing" effect in them. Aligning five identical probes over observatories on the North American continent has allowed scientists to collect coordinated measurements along the Earth's magnetic field lines, thereby providing the first comprehensive look at the onset of substorms and the manner in which they trigger auroral eruptions.

In 2009, NASA assigned two of the THEMIS satellites to a new mission—the Acceleration, Reconnection, Turbulence and Electrodynamics of the Moon's Interaction with the Sun (ARTEMIS) mission—to measure solar wind turbulence at the Moon. Having repositioned two of the five THEMIS probes in coordinated, lunar-equatorial orbits, ARTEMIS is now performing the first systematic, two-point observations of the distant magnetotail, the solar wind, and the lunar space environment. The primary objectives of the mission are to study how particles are accelerated at reconnection sites and shocks and how turbulence develops and evolves in Earth's magnetotail and in the solar wind. The mission will determine the structure, formation, refilling, and downstream evolution of the lunar wake. The three remaining THEMIS satellites continue to study substorms that are visible in the Northern Hemisphere as aurora borealis.

The German Aerospace Center (DLR) and the Technical University of Braunschweig developed and tested the fluxgate magnetometer (FGM), which is included on each probe. The Austrian Space Agency (ASA), the Space Research Institute of the Austrian Academy of Sciences (IWF/ÖAW), the Canadian Space Agency (CSA), and the French National Center for Space Studies (CNES) also contributed to this mission. NASA's contributions include the spacecraft, the launch, the ground network of observatories, and overall mission management.

For more information on THEMIS, please visit *http://www.nasa.gov/mission_pages/themis/main/index.html*.

For more information on ARTEMIS, please visit *http://www.nasa.gov/mission_pages/artemis/index.html*.

Science

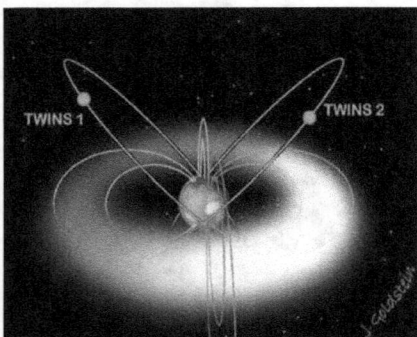

The highly elliptical orbit of TWINS offers a good view of the ring current—a hoop of charged particles that encircles Earth. (Credit: Southwest Research Institute)

Germany

TWINS
Two Wide-angle Imaging Neutral-atom Spectrometers

NASA's Two Wide-angle Imaging Neutral-atom Spectrometers (TWINS) mission is providing stereo imaging of the Earth's magnetosphere—the region surrounding the planet that is controlled by its magnetic field and contains the Van Allen radiation belts and other energetic charged particles. The two identical instruments, TWINS-A and TWINS-B, are providing three-dimensional (3-D) global visualization of this region, which is leading to greatly enhanced understanding of the connections between different regions of the magnetosphere and their relation to the solar wind. The German Aerospace Center (DLR) and the University of Bonn are collaborating with NASA on the TWINS mission.

The TWINS-A instrument was launched on a U.S. satellite in 2006, and the TWINS-B spacecraft launched on a U.S. satellite in 2008. With the two instruments in orbit, the TWINS mission began stereo imaging in 2008 and has since observed numerous co-rotating interaction region (CIR)–driven storms. This stereo imaging is yielding substantial progress in the characterization of the global 3-D distribution of the ring current.

The TWINS mission images the charge exchange neutral atoms over a broad energy range (~1–100 kiloelectron volts [keV]) using two identical instruments on two widely spaced high-altitude, high-inclination spacecraft. This enables 3-D visualization and the resolution of large-scale structures and dynamics within the magnetosphere. In contrast to traditional space experiments, which make measurements at only one point in space, imaging experiments provide simultaneous viewing of different regions of the magnetosphere. Stereo imaging as performed by TWINS takes the next step of producing 3-D images, which greatly improve our understanding of the global aspects of the terrestrial magnetosphere.

The TWINS instrumentation consists of an energetic neutral atom (ENA) imager and a Lyman alpha detector. The ENA imager provides indirect remote sensing of the ring current ions, and the Lyman alpha detector gives a measure of the neutral hydrogen cloud around the Earth, known as the geocorona. Additional environmental sensors are providing contemporaneous measurements of the local charged-particle environments. The University of Bonn provided the Lyman-alpha detector devices for the TWINS instruments and is participating in the joint analysis of the scientific data from the mission.

For more information on the TWINS mission, please visit *http://science.nasa.gov/missions/twins/*.

India

GPM/Megha-Tropiques
Global Precipitation Measurement/Megha-Tropiques

The Global Precipitation Measurement (GPM) mission is an international network of satellites that provide next-generation global observations of rain and snow. Building upon the success of the Tropical Rainfall Measuring Mission (TRMM), the GPM concept centers on the deployment of a "Core" satellite carrying an advanced radar/radiometer system to measure precipitation from space and serve as a reference standard to unify precipitation measurements from a constellation of research and operational satellites. The GPM Core Observatory carries the first space-borne Ku/Ka-band Dual-frequency Precipitation Radar (DPR), provided by the Japan Aerospace Exploration Agency (JAXA), as well as a multichannel GPM Microwave Imager (GMI) provided by NASA.

An artist's depiction of Megha-Tropiques satellite orbiting Earth. (Credit: French National Center for Space Studies)

Through improved measurements of precipitation globally, the GPM mission will help to advance our understanding of Earth's water and energy cycle, improve climate prediction, advance numerical weather prediction, improve forecasting of extreme events that cause natural hazards and disasters, and extend current capabilities in using accurate and timely information of precipitation to directly benefit society. GPM, initiated by NASA and JAXA as a global successor to TRMM, comprises a consortium of international space agencies, including the French National Center for Space Studies (CNES), the Indian Space Research Organisation (ISRO), the U.S. National Oceanic and Atmospheric Administration (NOAA), the European Organisation for the Exploitation of Meteorological Satellites (EUMETSAT), and others. The GPM Core Observatory launched in February 2014 on a Japanese H-IIA launch vehicle.

As a part of the future GPM constellation of satellites, NASA is partnering with ISRO and CNES on the joint Indian-French Megha-Tropiques mission, which was launched in 2011. Megha-Tropiques's main instruments are the Microwave Analysis and Detection of Rain and Atmospheric Structures (MADRAS) microwave imager for studying precipitation and cloud properties, the Sounder for Probing Vertical Profiles of Humidity (SAPHIR) microwave sounding instrument for atmospheric water vapor, and the Scanner for Radiation Budget (ScaRaB) radiometer devoted to the measurement of outgoing radiative fluxes at the top of the atmosphere.

NASA, ISRO, and CNES benefit from partnering on GPM and Megha-Tropiques through data sharing, scientific collaborations on satellite intercalibration, precipitation retrieval algorithm development, and ground validation. Additionally, MADRAS and SAPHIR data help NASA to improve sampling for GPM multisatellite global precipitation products.

For more information on the GPM mission, please visit *http://pmm.nasa.gov/precipitation-measurement-missions*.

Science

India

Oceansat-2

Oceansat-2, launched by India in 2009, is second in the Indian Space Research Organisation's (ISRO) series of Indian remote sensing satellites dedicated to ocean research (though Oceansat-2 also conducts atmospheric science). Oceansat-2 carries three sensors: the Ocean Color Monitor (OCM-2), which retrieves sea spectral reflectance; a Ku-band pencil-beam scatterometer capable of measuring ocean surface vector winds; and an Italian payload called the Radio Occultation Sounder for Atmosphere (ROSA), which provides a vertical profile of the atmosphere.

Oceansat-2 is providing continuity with the services and applications of the Oceansat-1 OCM-1 data and with additional data from a Ku-band pencil-beam scatterometer. Oceansat-2 is a three-axis, body-stabilized spacecraft placed into a near-circular Sun-synchronous orbit at an altitude of 720 kilometers, with an equatorial crossing time of around 1,200 hours.

NASA and ISRO are cooperating on Oceansat-2 by evaluating and optimizing OCM-2 data for research and increasing access to the global data. ISRO, NASA, and the U.S. National Oceanic and Atmospheric Administration (NOAA) share the common goal of optimizing the quality and maximizing the utility of the Oceansat-2 scatterometer data for the benefit of future global and regional scientific research and operational applications. The cooperative plan envisions continuing efforts to optimize the quality of the operational data stream over the near term and longer term for climate research.

For more information on Oceansat-2, please visit *http://www.isro.org/satellites/oceansat-2.aspx*.

Data from Oceansat-2's OCM-2 instrument was used to produce this image of India and the waters of the Bay of Bengal. (Credit: Indian Space Research Organisation)

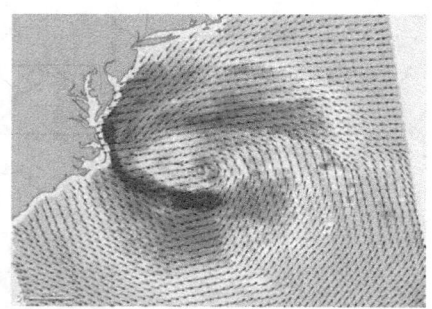

This map, produced with data from Oceansat-2's Ku-band pencil-beam scatterometer, shows the strength and direction of Hurricane Sandy's ocean surface winds in October 2012. (Credit: Indian Space Research Organisation/NASA)

Italy

BepiColombo

BepiColombo is a planned two-spacecraft mission consisting of the Mercury Planetary Orbiter (MPO), provided by the European Space Agency (ESA), and the Mercury Magnetospheric Orbiter (MMO), provided by the Japan Aerospace Exploration Agency (JAXA). The MPO will study the surface and internal composition of the planet, while the MMO will study Mercury's magnetosphere. NASA will contribute the Strofio instrument as part of the Italian Space Agency (ASI)–led Search for Exospheric Refilling and Emitted Natural Abundances (SERENA) suite of instruments on board the MPO.

An artist's impression of BepiColombo starting its 6-year interplanetary journey to Mercury. (Credit: European Space Agency)

Strofio is designed to obtain the first direct measurements of absolute abundances—both chemical and isotopic—of the exospheric constituents sampled along the low-altitude polar orbit of the MPO spacecraft. Strofio is the only instrument that can detect all of the neutral species that reach the spacecraft; optical and x-ray spectrometers can detect only those with emissions or absorptions at specific wavelengths. In contrast to optical techniques, Strofio can make measurements with high sensitivity, both in full sunlight and in Mercury's shadow. This unique capability enables Strofio to determine the diurnal and latitudinal variations in Mercury's exosphere that are critical to understanding its origins and variability. Strofio will operate with the other components of the SERENA instrument suite: the Emitted Low-Energy Neutral Atoms (ELENA) instrument, the Miniature Ion Precipitation Analyzer (MIPA), and the Planetary Ion Camera (PICAM).

The BepiColombo mission is scheduled to be launched by ESA in 2015 from Kourou, French Guiana.

For more information on BepiColombo and Strofio, please visit *http://sci.esa.int/bepicolombo* and *http://discovery.nasa.gov/strofio.cfml*.

The BepiColombo Mercury Composite Spacecraft is ready for a launcher separation shock test at ESA's European Space Research and Technology Centre (ESTEC) in Noordwijk, the Netherlands. (Credit: European Space Agency)

Science

Italy

Dawn Asteroid Rendezvous Mission

NASA's Dawn mission, launched in 2007, sent an orbiting space probe to examine the dwarf planet Ceres and the asteroid Vesta, two of the most massive members of the asteroid belt. In addition to being the first NASA science mission to make use of ion propulsion, Dawn will also be the first mission to visit and orbit two distinct bodies other than Earth and the Moon.

Dawn's goal is to characterize the conditions and processes of the solar system's earliest epoch by investigating in detail two of the largest protoplanets that have remained relatively intact since their formations. Together with many other smaller bodies, Ceres and Vesta reside in the extensive zone between Mars and Jupiter called the asteroid belt. Each has followed a very different evolutionary path, constrained by the diversity of processes that operated during the first few million years of solar system evolution.

The Dawn spacecraft successfully entered into orbit around Vesta in July 2011, and it has since revealed that the asteroid is a layered planetary building block with an iron core—the only one known to have survived the earliest days of the solar system. Dawn departed Vesta in September 2012 and will begin to study Ceres upon its arrival in 2015.

The Italian Space Agency (ASI) developed, supported, and integrated the Visual and Infrared Mapping Spectrometer (VIR-MS) instrument for the Dawn Mission. The VIR-MS is a high-spatial-resolution spectrometer that is used to study both the surface and atmosphere of the two protoplanets by acquiring a spectrally resolved image of a two-dimensional scene.

For more information on the Dawn mission, please visit *http://dawn.jpl.nasa.gov/mission/index.asp*.

An artist's concept of the Dawn spacecraft. (Credit: University of California, Los Angeles)

This image obtained by the framing camera on NASA's Dawn spacecraft shows the south pole of the giant asteroid Vesta. (Credit: NASA/Jet Propulsion Laboratory/University of California, Los Angeles/Max Planck Society/German Aerospace Center/Institute of Computer and Network Engineering, Braunschweig)

Italy

Fermi Gamma-ray Space Telescope

The Fermi Gamma-ray Space Telescope, launched by NASA in 2008, is a spacecraft with the ability to detect gamma rays created by the most energetic objects and phenomena in the universe. Among the topics of cosmological interest studied by the mission are dark matter and the periods of star and galaxy formation in the early universe. The mission is an astrophysics and particle physics partnership developed by NASA in collaboration with the U.S. Department of Energy, with important contributions from academic institutions and partners in France, Germany, Italy, Japan, Sweden, and the United States.

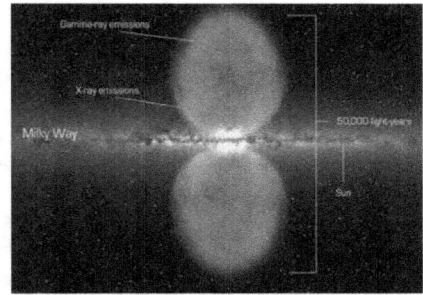

The Fermi Gamma-ray Space Telescope discovered gamma-ray bubbles that extend 50,000 light-years—or roughly half of the Milky Way's diameter.

Because of their tremendous energy, gamma rays travel through the universe largely unobstructed. This means that Fermi is able to observe gamma-ray sources located near the edge of the visible universe. Gamma rays detected by Fermi originate near the otherwise-obscured central regions of exotic objects like supermassive black holes, pulsars, and gamma-ray bursts.

In order to study these high-energy waves, two main instruments are used: the Large Area Telescope (LAT) and the Gamma-ray Large Area Space Telescope (GLAST) Burst Monitor (GBM). When in survey mode, the LAT scans the entire sky every 3 hours. The instrument detects photons with energies ranging from 20 million electron volts to over 300 billion electron volts. In its first month of operation, the GBM instrument spotted 31 gamma-ray bursts, which occur when massive stars die or when orbiting neutron stars spiral together and merge. By the end of 2011, Fermi had detected more than 100 gamma-emitting pulsars; before its launch, only 7 of these were known to emit gamma rays. Newly developed methods of analysis promise to spur additional discoveries.

The Italian Space Agency (ASI) provided particle physics and astrophysics expertise toward the mission and built the LAT Tracker. Along with ASI and NASA, the following organizations are also contributors to the Fermi mission: the German Aerospace Center (DLR), the French Alternative Energies and Atomic Energy Commission (CEA), Japan Aerospace Exploration Agency (JAXA), and the Swedish Royal Institute of Technology (KTH).

For more information on the FERMI mission, please visit *http://fermi.gsfc.nasa.gov/*.

Science

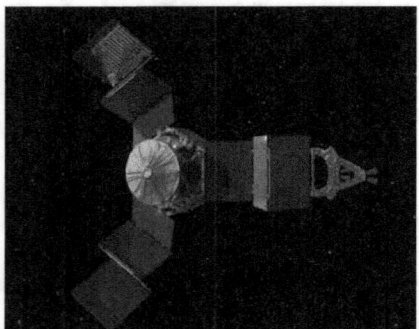

An artist's rendering of the Juno spacecraft.

Italy

Juno

NASA's Juno spacecraft, launched in 2011 and slated to arrive in orbit around Jupiter in 2016, will gather data on Jupiter's interior structure, atmospheric composition and dynamics, and polar magnetosphere. This data will be used by scientists to study the origin and evolution of Jupiter, thereby furthering our understanding of planetary and solar-system formation.

Using a spinning, solar-powered spacecraft, Juno will make maps of the gravitational and magnetic fields, auroral emissions, and atmospheric composition of Jupiter from a polar orbit. Juno will carry precise high-sensitivity radiometers, magnetometers, and gravity science systems. Juno's 32 orbits over 11 days will sample Jupiter's full range of latitudes and longitudes. From its polar perspective, Juno will combine in situ and remote sensing observations to explore the polar magnetosphere and determine what drives Jupiter's remarkable auroras. Jupiter's solid core and abundance of heavy metals in the atmosphere make it an ideal model to help improve our understanding of the origin of giant planets. Juno will measure global abundances of oxygen and nitrogen by mapping the gravitational field and using microwave observations of water and ammonia.

The Italian Space Agency (ASI) contributed two instruments for the Juno mission: the Jovian Infrared Auroral Mapper (JIRAM) and the Ka-band Transponder (KaT). The JIRAM instrument will be used to collect high-resolution images of Jupiter's atmosphere, combining the use of an infrared camera and a spectrometer. KaT will be used to aid in the sensitive gravity tracking experiments. ASI also contributed a novel education and outreach article: a small aluminum plaque honoring Galileo Galilei. ASI intends to use this item to create new interest in science by highlighting its own heritage.

For more information on the Juno mission, please visit *http://juno.nasa.gov/* and *http://missionjuno.swri.edu/*.

Italy

MRO
Mars Reconnaissance Orbiter

The Italian Space Agency (ASI) provided the Shallow Subsurface Radar (SHARAD), one of the six science instruments on NASA's Mars Reconnaissance Orbiter (MRO) mission, which launched in August 2005 and arrived at Mars in March 2006.

MRO is investigating the history of water on Mars by performing remote sensing science investigations from Mars orbit. MRO conducts observations in several parts of the electromagnetic spectrum—including ultraviolet and visible imaging, visible to near-infrared imaging spectrometry, thermal infrared atmospheric sounding, and radar subsurface profiling—at spatial resolutions that are substantially better than any preceding Mars orbiter. MRO is also used to identify and characterize sites for future landed missions and to provide critical telecommunications relay capability for other Mars missions.

This artist's concept of the Mars Reconnaissance Orbiter highlights the spacecraft's radar capability. The SHARAD instrument "looks" under the first few hundred feet (up to 1 kilometer) below the Martian surface. (Credit: NASA/Jet Propulsion Laboratory)

SHARAD is probing the three-dimensional cross-sections of the layer structure of the upper 500–1,000 meters of the Martian surface with high-vertical-resolution and landform-scale horizontal sampling for the shallow subsurface. The goal is to characterize subsurface layering and to determine whether liquid or frozen water exists within the first few hundred feet of Mars's crust. SHARAD works by sending out and recording the return of radar pulses of energy, emitted and captured by the SHARAD antenna. The timing of the radar return is sensitive to changes in the electrical properties of the rock, sand, and any water present in the surface and subsurface. Water, like high-density rock, is very conducting and will have a very strong radar return. Changes in the reflection characteristics of the subsurface, such as layers deposited by geological processes throughout Mars's history, will also be visible.

For more information on MRO and SHARAD, please visit *http://marsprogram.jpl.nasa.gov* and *http://www.nasa.gov/mission_pages/MRO/spacecraft/sc-instru-sharad.html*.

The Mars Reconnaissance Orbiter's SHARAD instrument returned this enhanced-color view of the eastern rim and floor of Victoria Crater on Mars, which the Mars Exploration Rover Opportunity was investigating at the same time.

Human Exploration and Operations

The former Leonardo MPLM arrives via STS-135 at the International Space Station (ISS), where it takes up residence as a permanent ISS module, adopting the new name of Permanent Multipurpose Module (PMM). The PMM adds an additional 2,472 cubic feet of stowage and science activity space to the ISS.

STS-135 mission specialist Sandy Magnus is surrounded by supplies and equipment in the PMM.

Italy

MPLM
Multi-Purpose Logistics Module

The Italian Space Agency's (ASI) contribution to the International Space Station (ISS) Program consisted of three Multi-Purpose Logistics Modules (MPLMs) named Leonardo, Raffaello, and Donatello. The Leonardo eventually became a Permanent Multipurpose Module (PMM). In exchange for the development and manufacturing of these elements as NASA-provided contributions to the ISS Program, NASA is providing to ASI several astronaut flight opportunities and utilization rights aboard the ISS.

NASA's contributions to the ISS Program include overall management, coordination, and integration of the ISS Program across the partnership for operations, safety, transportation, and integrated hardware performance. NASA hardware components include the U.S. laboratory module, the airlock for extravehicular activities (EVA), truss segments to support the U.S. solar arrays and thermal radiators, three connecting nodes, and living quarters. NASA also provides cargo transportation via the Commercial Resupply Services (CRS) contract.

The ISS functions as an orbital microgravity and life sciences laboratory, a test bed for new technologies in areas such as life support and robotics, and a platform for astronomical and Earth observations. The ISS also serves as a unique engineering test bed for flight systems and operations critical to NASA's future exploration missions. U.S. research on the ISS concentrates on the long-term effects of space travel on humans, as well as on technology development activities in support of exploration. Hailed as one of the most ambitious engineering feats in human history, the ISS is a stepping-stone for human exploration and scientific discovery beyond low-Earth orbit.

The MPLMs were pressurized, reusable logistics modules that functioned both as cargo carriers between the ground and the ISS and as attached ISS modules. As cargo carriers aboard the Space Shuttles, MPLMs carried laboratory racks filled with equipment, experiments, and supplies to the ISS, then carried excess materials from the ISS to the ground. While berthed to the ISS, the MPLMs were fully functioning extra work areas (with components that provided life support, fire detection and suppression, electrical distribution, and computer functions).

During Space Shuttle Discovery's final flight in 2011, a refurbished Leonardo was delivered and permanently mounted to the ISS. This former MPLM, now a PMM, is berthed to the nadir port of Node 1. The PMM added an additional 2,472 cubic feet of stowage and science activity space to the ISS.

For more information on ASI's contributions to the ISS, please visit *http://mplm.msfc.nasa.gov/mission.html* and *http://www.nasa.gov/mission_pages/station/behindscenes/PMM_transformation.html*.

Italy

NuSTAR
Nuclear Spectroscopic Telescope Array

NASA's Nuclear Spectroscopic Telescope Array (NuSTAR) mission, launched in June 2012, will allow astronomers to study the universe in high-energy x rays. Nine days after its launch, engineers at NuSTAR's mission control at the University of California Berkeley sent a signal to the spacecraft to begin deploying its 10-meter mast. The mast was deployed and secured over a span of 26 minutes, making it the first deployed mast ever used on a space telescope. As a result of this achievement, NuSTAR became the first focusing hard x-ray telescope to orbit Earth, and it is expected to greatly exceed the performance of the largest ground-based observatories that have observed this region in the electromagnetic spectrum. NuSTAR will also complement astrophysics missions that explore the cosmos in other regions of the spectrum. NASA is collaborating with the Italian Space Agency (ASI) and Danish Technical University on the NuSTAR mission.

NuSTAR is able to focus the high-energy x-ray light into sharp images because of a complex, innovative telescope design. High-energy light is difficult to focus because it reflects off mirrors only when hitting at nearly parallel angles. NuSTAR ingeniously solves this problem by using nested shells of mirrors in its design. It has the most nested shells ever used in a space telescope—133 in each of two optic units. The mirrors were molded from ultrathin glass similar to that found in laptop screens and glazed with even thinner layers of reflective coating.

ASI is providing the mission with use of the Malindi Ground Station, operated by ASI and located in Malindi, Kenya, as the primary NuSTAR data downlink and command uplink facility.

For more information on NuSTAR, please visit *http://www.nasa.gov/nustar*.

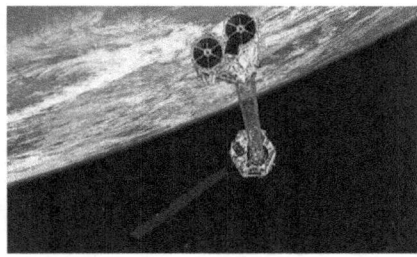

An artist's concept of NASA's NuSTAR spacecraft orbiting Earth; NuSTAR hunts for hidden black holes and other exotic cosmic objects. (Credit: NASA/Jet Propulsion Laboratory–Caltech)

This view of the historical supernova remnant Cassiopeia A was taken by NASA's NuSTAR spacecraft. Blue indicates the highest energy x-ray light, where NuSTAR has made the first resolved image ever of this source. Red and green show the lower end of NuSTAR's energy range, which overlaps with NASA's high-resolution Chandra X-ray Observatory. The starry background picture is from the Digitized Sky Survey. (Credit: NASA/Jet Propulsion Laboratory–Caltech/Digitized Sky Survey)

Science

Italy

Swift Gamma Ray Burst Explorer

NASA's Swift Gamma Ray Burst Explorer mission is a multiwavelength astrophysics observatory that is making a comprehensive study of hundreds of gamma-ray bursts (GRBs) in order to determine the origin and physical processes of GRB events. The Swift mission was launched by NASA in November 2004 and was expected to have an orbital lifetime of approximately 8 years. Updated orbital lifetime projections for Swift predict that it will remain in orbit potentially until 2022. The Italian Space Agency (ASI) and the United Kingdom Space Agency are cooperating with NASA on the operation of the Swift mission.

GRBs are incredibly intense releases of gamma radiation, which is a particularly energetic form of light that can be generated only by the most powerful astronomical events. Swift is a next-generation satellite that observes GRBs—a first-of-its-kind, multi-wavelength observatory dedicated to advancing GRB science. Its three instruments work together to patrol about one-sixth of the sky at a time in order to pinpoint and observe GRBs and afterglows in the gamma-ray, x-ray, ultraviolet, and optical wavebands.

ASI provided the Swift x-ray mirror and associated components for the x-ray telescope (XRT), and it also led the development of XRT data-analysis software and data-processing software. The XRT takes images and measures the light energy (spectra) of afterglows that follow GRBs. Additionally, the Italian ground station at Malinda, Kenya, serves as the primary ground station for the entire Swift mission. ASI also provided the network infrastructure (ASINet) and maintains the transmission of Swift data between Malinda and the ASINet gateway at NASA's Johnson Space Center (JSC). ASI also provides the Italian Swift Archive Center.

For more information on Swift and the XRT, please visit *http://swift.gsfc.nasa.gov* and *http://www.nasa.gov/mission_pages/swift/spacecraft/index.html*.

An artist's rendering of the Swift spacecraft with a gamma-ray burst in the background. (Credit: Spectrum Astro)

This stunning vista of the Andromeda Galaxy at ultraviolet wavelengths was recorded by NASA's Swift satellite. The mosaic image is composed of 330 individual images covering a region 200,000 light-years wide.

Japan

Astro-H

Astro-H is a powerful orbiting observatory being developed by the Japan Aerospace Exploration Agency (JAXA) for the purpose of studying extremely energetic processes in the universe. Astro-H is expected to launch in 2014 from the Tanegashima Space Center (TNSC) in Japan. The Astro-H mission is a cooperative Japan-U.S. x-ray astronomy mission initiated by JAXA's Institute of Space and Astronautical Science (ISAS). Astro-H will provide unprecedented sensitivity for high-resolution spectroscopy of cosmic x-ray sources.

An artist's rendering of ASTRO-H spacecraft with galaxies—one of the things it will study—in the background. (Credit: JAXA/Akihiro Ikeshita)

Astro-H will be comprised of a suite of four highly complementary instruments spanning the x-ray energy band from 0.3 to 600 kiloelectron volts (keV). The Soft X-ray Spectrometer (SXS), which is being developed jointly by a team led by NASA's Goddard Space Flight Center (GSFC) and JAXA's ISAS, is a high-resolution, nondispersive x-ray spectrometer operating between 0.3 and 12 keV. It represents the core instrument on Astro-H, providing a high-resolution spectroscopic capability for the Astro-H mission and covering the energy band where all of the astrophysically abundant elements have characteristic emission lines that can be used for a wide range of spectral studies of matter under extreme conditions. Three additional scientific instruments provided by JAXA will extend the bandpass to produce an observatory with extraordinary new capabilities. The Soft X-ray Imager (SXI) will cover the same energy band as the SXS and will expand the observatory's field of view with a new-generation charge-coupled device (CCD) camera. The Hard X-ray Imager (HXI) will perform sensitive imaging spectroscopy in the 5–80 keV band using specially coated x-ray optics, and the nonimaging Soft Gamma-ray Detector (SGD) will extend the observatory's energy band to 600 keV.

The SXS will test theories of structure formation by measuring the velocity field of x-ray–emitting gas in clusters of galaxies and the energy output from the jets and winds of active galaxies. SXS will measure metal abundances in the oldest galaxies, providing information about the origin of the elements. SXS will also observe matter in extreme gravitational fields, obtaining time-resolved spectra from material approaching the event horizon of a black hole.

For more information on Astro-H, please visit *http://science.nasa.gov/missions/astro-h*.

Human Exploration and Operations

CALET will be located on the International Space Station's Japanese Experiment Module–Exposed Facility.

Japan

CALET
CALorimetric Electron Telescope

The CALorimetric Electron Telescope (CALET) is an all-sky gamma-ray and electron astrophysics observatory selected by the Japan Aerospace Exploration Agency (JAXA) for deployment on the Japanese Experiment Module (JEM)–Exposed Facility (EF) on the International Space Station (ISS). CALET is planned for a 2014 launch on an H-II Transfer Vehicle (HTV) to the ISS.

The goal of the CALET Mission is to investigate the high-energy universe by observing cosmic-ray electrons. The CALET detector consists of the Imaging Calorimeter (IMC), the Total AbSorption Calorimeter (TASC), the Charge Detector (CHD), and a gamma-ray burst monitor. CALET has a unique capability to observe electrons and gamma rays in the high-energy spectrum. This capability enables CALET to search for nearby cosmic-ray sources and dark matter. CALET will also monitor gamma-ray bursts.

JAXA is developing the equipment and most mission instrumentation for the satellite bus, providing the launch on the HTV to the ISS, and providing the on-orbit location for CALET on board the JEM-EF. JAXA will also operate CALET in orbit and has established the international CALET Science Team. NASA is providing technical expertise in the design, development, and testing of CALET mission equipment. NASA has also selected U.S. scientists to participate on the CALET Science Team, and NASA will archive CALET data at NASA's Goddard Space Flight Center (GSFC).

For more information on the CALET mission, please visit *http://calet.phys.lsu.edu/Instrumentation.php* and *http://asd.gsfc.nasa.gov/*.

Japan

DTN
Disruption Tolerant Networking

Disruption Tolerant Networking (DTN) is a new technology that will extend the reach of the terrestrial Internet across the solar system. DTN addresses the common spaceflight configuration in which data must be exchanged between Earth and remote space vehicles but for which there is no end-to-end data path currently available, a circumstance that can cause long delays.

JAXA's Kibo module will provide networks for two-way connectivity from JAXA's ground facilities to a NASA payload.

NASA is currently partnering with multiple space agencies to develop, demonstrate, and infuse the DTN technology to support international cooperative space exploration missions launching in 2015 and beyond. Part of NASA's risk-mitigation strategy is to flight-test the DTN protocols in a real mission environment. The ISS provides an ideal platform, and NASA has begun to use the ISS as an international flight laboratory for maturing DTN to flight readiness.

Japan Aerospace Exploration Agency's (JAXA) participation in the DTN development program brings many benefits, including the use of JAXA's Data Relay Test Satellite (DRTS) and the employment of the Japanese Experiment Module on board the ISS (called the JEM, or "Kibo") to expand the topology for testing and flight-demonstrating the new protocols in a collaborative international environment, leading to increased future interoperability for space exploration missions.

JAXA will use the DTN protocols to interact with special software test applications that will be implemented in the NASA Commercial Generic Bioprocessing Apparatus (CGBA) payload located in NASA's Destiny Laboratory on the ISS, which is equipped with DTN capabilities. The two CGBA—CGBA-4 and CGBA-5—serve as communications test computers that transmit messages between the ISS and ground mission-control centers.

Two-way connectivity will be established from JAXA's ground facilities to NASA via the JAXA DRTS and the ISS onboard payload networks (including networks within the Kibo module) in order to establish an internetworked data communications dialog. This cooperation is built upon the mutual benefit of joint testing, evaluation, and demonstration of interoperable space internetworking technologies utilizing the emerging suite of DTN data communications protocols.

For more information on Disruption Tolerant Networking, please visit *http://www.nasa.gov/mission_pages/station/research/experiments/DTN.html*.

Science

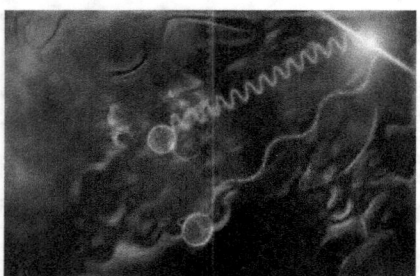

In this illustration, one photon (purple) carries a million times the energy of the other. Some theorists predict travel delays for higher-energy photons; however, Fermi data on two photons from a gamma-ray burst fail to show this effect. (Credit: NASA/Sonoma State University/Aurore Simonnet)

Japan

Fermi Gamma-ray Space Telescope

The Fermi Gamma-ray Space Telescope, launched by NASA in 2008, is a spacecraft with the ability to detect gamma rays created by the most energetic objects and phenomena in the universe. Among the topics of cosmological interest studied by the mission are dark matter and the periods of star and galaxy formation in the early universe. The mission is an astrophysics and particle physics partnership developed by NASA in collaboration with the U.S. Department of Energy, with important contributions from academic institutions and partners in France, Germany, Italy, Japan, Sweden, and the United States.

Because of their tremendous energy, gamma rays travel through the universe largely unobstructed. This means that Fermi is able to observe gamma-ray sources located near the edge of the visible universe. Gamma rays detected by Fermi originate near the otherwise-obscured central regions of exotic objects like supermassive black holes, pulsars, and gamma-ray bursts.

In order to study these high-energy waves, two main instruments are used: the Large Area Telescope (LAT) and the Gamma-ray Large Area Space Telescope (GLAST) Burst Monitor (GBM). When in survey mode, the LAT scans the entire sky every 3 hours. The instrument detects photons with energies ranging from 20 million electron volts to over 300 billion electron volts. In its first month of operation, the GBM instrument spotted 31 gamma-ray bursts, which occur when massive stars die or when orbiting neutron stars spiral together and merge. By the end of 2011, Fermi had detected more than 100 gamma-emitting pulsars; before its launch, only 7 of these were known to emit gamma rays. Newly developed methods of analysis promise to spur additional discoveries.

NASA is working with a Japan Aerospace Exploration Agency (JAXA)–led team comprising partners from the University of Tokyo, Tokyo Institute of Technology, Institute for Cosmic-Ray Research (ICRR), JAXA's Institute of Space and Astronautical Science (ISAS), and Hiroshima University on the Fermi mission. The Japanese partners provided oversight in the making of the silicon-strip detectors in the LAT Tracker. Along with Japan and NASA, the following organizations are also contributors to the Fermi mission: the German Aerospace Center (DLR), Italian Space Agency (ASI), the French Alternative Energies and Atomic Energy Commission (CEA), and the Swedish Royal Institute of Technology (KTH).

For more information on the FERMI mission, please visit *http://fermi.gsfc.nasa.gov/*.

Japan

GPM
Global Precipitation Measurement

The Global Precipitation Measurement (GPM) mission is an international network of satellites that provide next-generation global observations of rain and snow. Building upon the success of the Tropical Rainfall Measuring Mission (TRMM), the GPM concept centers on the deployment of a "Core" satellite carrying an advanced radar/radiometer system to measure precipitation from space and serve as a reference standard to unify precipitation measurements from a constellation of research and operational satellites. The GPM Core Observatory, launched in February 2014, carries the first space-borne Ku/Ka-band Dual-frequency Precipitation Radar (DPR) provided by the Japan Aerospace Exploration Agency (JAXA), as well as a multichannel GPM Microwave Imager (GMI) provided by NASA.

Through improved measurements of precipitation globally, the GPM mission will help to advance our understanding of Earth's water and energy cycle, improve climate prediction, advance numerical weather prediction, improve forecasting of extreme events that cause natural hazards and disasters, and extend current capabilities in using accurate and timely information of precipitation to directly benefit society. GPM, initiated by NASA and JAXA as a global successor to TRMM, comprises a consortium of international space agencies, including the French National Center for Space Studies (CNES), the Indian Space Research Organisation (ISRO), the U.S. National Oceanic and Atmospheric Administration (NOAA), the European Organisation for the Exploitation of Meteorological Satellites (EUMETSAT), and others. The GPM Core Observatory launched in February 2014 on a Japanese H-IIA launch vehicle.

JAXA delivered the DPR to NASA in March 2012. The DPR was designed and built by JAXA and Japan's National Institute of Information and Communications Technology (NICT). The DPR is more sensitive to light rain rates and snowfall than the TRMM precipitation radar. In addition, simultaneous measurements by the overlapping of the DPR's Ka/Ku-bands can provide new information on particle drop size distributions over moderate precipitation intensities. By providing new microphysical measurements from the DPR to complement cloud and aerosol observations, GPM is expected to provide further insights into how precipitation processes may be affected by human activities.

For more information on the GPM mission, please visit *http://pmm.nasa.gov/precipitation-measurement-missions*.

The Dual-frequency Precipitation Radar (DPR) instrument's electrical integration onto the GPM Core Observatory was completed in May 2012 at NASA's Goddard Space Flight Center. The DPR is comprised of two radars: Ka-band Radar (13 gigahertz) and Ku-band Radar (35 gigahertz). JAXA built both radars for the GPM Core Observatory.

Science

An artist's rendering of Hayabusa above the surface of the Itokawa asteroid.

Hayabusa's sample-return capsule and parachute lie on the ground in Australia's Woomera Prohibited Area. (Credit: Japan Aerospace Exploration Agency)

Japan

Hayabusa/Hayabusa-2

Hayabusa was an unmanned mission led by the Japan Aerospace Exploration Agency (JAXA) that collected a surface sample of material from the asteroid 25143 Itokawa and returned the sample to Earth for analysis.

Hayabusa was launched in May 2003 and rendezvoused with the Itokawa asteroid in September 2005. Hayabusa mapped the asteroid in the visible, infrared, and x-ray spectra while also performing gravity modeling with its laser altimeter.

After Hayabusa collected samples from the asteroid, the samples were stored in containers inside a sealed sample return capsule that detached from the Hayabusa spacecraft, entered Earth's atmosphere, and used a parachute for a soft landing on the ground in Australia in June 2010.

The Hayabusa mission tested four advanced technology systems: the electric propulsion (ion drive) engines, an autonomous navigation system, the sample collection system, and the sample capsule that re-entered the Earth's atmosphere. NASA resources provided scientific support and backup spacecraft tracking, telemetry, command, and navigation support through the Deep Space Network.

The Hayabusa mission continues to advance our understanding of near-Earth asteroids. NASA scientists are participating in experiments using the returned asteroid material samples.

In 2014, JAXA plans to launch Hayabusa-2 to the asteroid 1999 JU3 in an attempt to build upon the success of the original Hayabusa mission, and NASA intends to participate in this follow-up by providing a level of support similar to what it provided during the original mission.

For more information on Hayabusa, please visit *http://hayabusa.jaxa.jp/e/*, *http://www.jspec.jaxa.jp/e/activity/hayabusa2.html*, and *http://nssdc.gsfc.nasa.gov/database/MasterCatalog?sc=2003-019A*.

Japan

Hinode

Hinode is a Japan Aerospace Exploration Agency (JAXA)–led solar physics mission that is providing the first solar optical telescope in space. NASA and the United Kingdom are collaborators on the mission. Launched by JAXA in 2006 from the Uchinoura Space Center in Japan, Hinode data is helping scientists to better understand how magnetic fields interact with plasma to produce solar variability and how the solar photosphere and corona act as a system. The Hinode mission aims to solve how the generation of the solar-magnetic field and its emergence through the photosphere governs the structure of the entire solar atmosphere.

An artist's concept of the Hinode spacecraft orbiting Earth with an active Sun in the background. (Credit: Japan Aerospace Exploration Agency)

The Hinode mission includes a suite of three science instruments—the Solar Optical Telescope (SOT), the x-ray telescope (XRT), and the Extreme-ultraviolet Imaging Spectrometer (EIS)—to study the interaction between the Sun's magnetic field and its high-temperature, ionized atmosphere. Hinode is circling Earth in a polar, Sun-synchronous orbit that allows the spacecraft's instruments to remain in continuous sunlight for 9 months each year.

Hinode's instruments have produced fantastic detail of both visible and magnetic features on the Sun's surface and in its atmosphere, the corona. Since 2006, the Hinode mission has helped to explain the origin of the solar wind, discovered potential candidates for how the corona gets so hot, and provided images of the complex magnetic structures looping up and out of active regions on the Sun. Hinode's Solar Optical Telescope has delivered images that show greatly magnified views of the Sun's surface. These images are revealing new details about solar convection, the process that drives the rising and falling of gases in the lowest atmospheric region, the photosphere. Hinode's Solar Optical Telescope is the first space-borne instrument to measure the strength and direction of the Sun's magnetic field.

Hinode captured this image of the January 6, 2011, solar eclipse. (Credit: Japan Aerospace Exploration Agency)

JAXA is the overall lead for the Hinode mission and provided the spacecraft, launch vehicle, and management of space operations. The satellite systems were developed by JAXA and Mitsubishi Electric Corporation. The large Solar Optical Telescope was jointly developed by the United States and Japan. JAXA and the National Astronomical Observatory of Japan (NAOJ) worked on the telescope optics, and NASA developed the focal-plane package. For the x-ray telescope, NASA provided grazing-incidence mirror optics, while JAXA provided the charge-coupled device (CCD) camera. Development of the EIS was led by the United Kingdom, with the support of NASA and JAXA.

For more information on Hinode, please visit *http://www.nasa.gov/mission_pages/solar-b/*.

Human Exploration and Operations

JAXA astronaut Akihiko Hoshide works in the newly installed Kibo Japanese Pressurized Module.

The International Space Station's Canadarm2 unberths the unpiloted JAXA HTV-3, filled with trash and unneeded items, in preparation for its release from the station.

Japan

International Space Station

The Japan Aerospace Exploration Agency's (JAXA) primary contributions to the International Space Station (ISS) Program are the Japanese Experiment Module (JEM)—called "Kibo," which means "hope" in Japanese—and the H-II Transfer Vehicle (HTV), an unmanned logistics resupply vehicle.

The ISS Program is a partnership among the space agencies of Canada, Europe, Japan, Russia, and the United States. International crews have continuously inhabited the ISS since November 2000. Cargo transportation is provided by European, Russian, and Japanese vehicles and by NASA's contracts for Commercial Resupply Services (CRS). Crew transportation and rescue is provided by the Russian Federal Space Agency and, beginning in 2017, by NASA's commercial crew transportation services.

NASA's contributions to the ISS Program include overall management, coordination, and integration of the ISS Program across the partnership for operations, safety, transportation, and integrated hardware performance. NASA hardware components include the U.S. laboratory module, the airlock for extravehicular activities (EVA), truss segments to support the U.S. solar arrays and thermal radiators, three connecting nodes, and living quarters.

The ISS functions as an orbital microgravity and life sciences laboratory, a test bed for new technologies in areas such as life support and robotics, and a platform for astronomical and Earth observations. The ISS also serves as a unique engineering test bed for flight systems and operations critical to NASA's future exploration missions. U.S. research on the ISS concentrates on the long-term effects of space travel on humans and on technology development activities in support of exploration. The ISS is a stepping-stone for human exploration and scientific discovery beyond low-Earth orbit.

The JEM laboratory houses facilities that support space medicine, biology, Earth observations, material production, and biotechnology and communications research. JEM experiments and systems are operated from the mission control room at Tsukuba Space Center (TKSC) in Japan.

The HTV is an autonomous logistical resupply vehicle that berths to the ISS using the Space Station Remote Manipulation System (SSRMS). The HTV can carry cargo in both its internal pressurized carrier as well as in an unpressurized carrier for the exterior placement of cargo such as batteries. It is launched on JAXA's H-IIB launch vehicle and can carry dry cargo, gases, and water. After delivering its cargo to the ISS, the HTV is reloaded with trash and waste products before it is unberthed and deorbited, whereupon it incinerates during reentry.

For more information on the ISS and Japan's contributions, please visit *http://iss.jaxa.jp/en/* and *http://www.nasa.gov/mission_pages/station/cooperation/index.html*.

Japan

MMS
Magnetospheric MultiScale

NASA's Magnetospheric MultiScale (MMS) mission is a solar-terrestrial probe mission comprising four identically instrumented spacecraft that will use Earth's magnetosphere as a laboratory to study the microphysics of three fundamental plasma processes: magnetic reconnection, energetic particle acceleration, and turbulence. These processes occur in all astrophysical plasma systems but can be studied in situ only in our solar system and most efficiently only in Earth's magnetosphere, where they control the dynamics of the geospace environment and play an important role in the processes known as "space weather." The four MMS spacecraft are currently planned for launch by NASA in 2014.

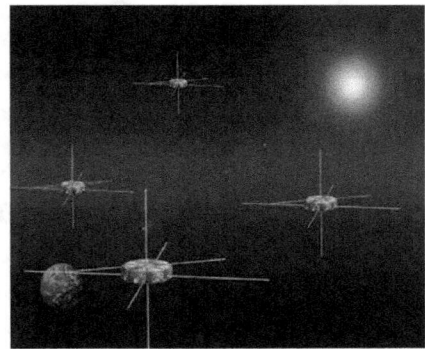

An artist's rendering of the four identical MMS spacecraft. (Credit: Southwest Research Institute)

NASA is collaborating on the MMS mission with the Japan Aerospace Exploration Agency (JAXA). NASA is also cooperating with the Aeronautics and Space Agency (ALR) of the Austrian Research Promotion Agency (FFG), the French National Center for Space Studies (CNES), and the Swedish National Space Board (SNSB).

MMS investigates how the magnetic fields of the Sun and Earth connect and disconnect, explosively transferring energy from one to the other in a process known as magnetic reconnection. This process limits the performance of fusion reactors and is the final governor of geospace weather, which affects modern technological systems such as telecommunications networks, Global Positioning System (GPS) navigation, and electrical power grids. MMS's four spacecraft will measure plasmas, fields, and particles in a near-equatorial orbit that will frequently encounter reconnection in action.

The four MMS spacecraft will carry identical instrument suites of plasma analyzers, energetic particle detectors, magnetometers, and electric field instruments; they will also carry a device to prevent spacecraft charging from interfering with the highly sensitive measurements required in and around the diffusion regions. JAXA is supporting the design and development of the Dual Ion Spectrometers (DIS) as part of the Fast Plasma Instrument (FPI) for the four MMS spacecraft. NASA is providing the spacecraft, the launch, and overall mission management.

For more information on MMS, please visit *http://mms.gsfc.nasa.gov/*.

Aeronautics Research

Japan

Sonic Boom Research

This F-18 aircraft from NASA's Dryden Research Center conducts sonic boom testing for the NASA–JAXA cooperation.

The Vibro-Acoustic Device is used in the NASA–JAXA sonic boom research testing.

A sonic boom is the thunder-like noise that a person on the ground hears when an aircraft flies overhead at supersonic speeds—faster than the speed of sound. Because the aircraft is traveling faster than sound, the pressure change in the air caused by its passage is nearly instantaneous. This shock wave of pressure travels out from the aircraft in all directions, eventually reaching the ground, where it is heard as a boom. The shape, length, and weight of the aircraft all affect the magnitude of the boom.

NASA and the Japan Aerospace Exploration Agency (JAXA) are working together to better understand the impact that sonic booms have on residential and commercial buildings and the resulting human response to the noise heard inside these buildings. NASA and JAXA are sharing test methods and the results of testing conducted by both agencies in the laboratory and in the field. Laboratory tests of human response to sonic booms indoors have been conducted in unique boom-simulation facilities at both NASA and JAXA, and the results have been shared to aid in the development of prediction models. In the field, JAXA has partnered with NASA for tests at Edwards Air Force Base in California of sonic boom transmission into large buildings. NASA and JAXA have also shared data from separate field tests of sonic boom propagation effects. Over the longer term, NASA and JAXA hope to be able to use these experiences and shared data, along with advances in sonic boom simulation, to predict human response to sonic booms experienced inside residential and commercial buildings.

NASA and JAXA are also cooperating in sonic boom modeling, which is a key technology that will contribute to the ability of next-generation supersonic aircraft to fly at supersonic speeds over land without significant disturbance to the public. Such a vehicle could fly from Los Angeles to New York City in less than 3 hours at Mach 1, and from Los Angeles to Tokyo in about 5 hours at Mach 2.

For more information on sonic booms, please visit *http://www.nasa.gov/centers/dryden/news/FactSheets/FS-016-DFRC.html*.

Japan

Suzaku

The Japan Aerospace Exploration Agency's (JAXA) Suzaku satellite, launched by JAXA in 2005, is a powerful orbiting observatory for studying extremely energetic processes in the universe, such as neutron stars, active and merging galaxies, black holes, and supernovae.

The Suzaku spacecraft features the X-ray Imaging Spectrometer (XIS), which is comprised of four individual charge-coupled device (CCD) x-ray cameras that provide high-sensitivity imaging over a large field of view and a hard x-ray detector (HXD) that is used for broadband spectroscopy up to the gamma-ray region. The Suzaku spacecraft also featured the high-resolution x-ray spectrometer (XRS); however, this instrument, which thermally detected individual x-ray photons and measured their energies with a high degree of precision and sensitivity, was only operational for 3 weeks due to a series of events associated with helium gas entering the dewar vacuum space. NASA collaborated with JAXA on the development of the XRS and XIS instruments.

Suzaku has been providing insight on galaxy clusters, which are millions of light-years across. Theses clusters can be examined using x-ray imaging due to the hot x-ray–emitting gas that fills the space between the galaxies. In 2011, Suzaku provided the clearest picture to date of the size, mass, and chemical content of the Perseus Galaxy Cluster, which is located 250 million light-years away. As scientists increase their understanding of the content of normal matter in galaxy clusters, they improve their ability to study the evolution of the universe.

For additional information on the Suzaku mission, please visit *http://www.nasa.gov/astro-e2* and *http://suzaku.gsfc.nasa.gov/*.

The Suzaku X-Ray Spectrometer in the laboratory.

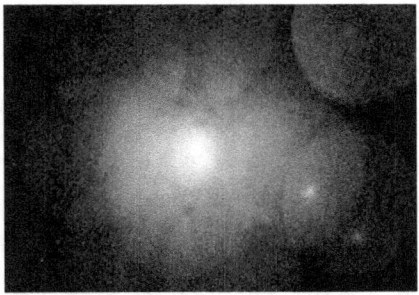

This image from Japan's Suzaku spacecraft shows the x-ray glow of the 100-million-degree-Fahrenheit gas that fills the Perseus cluster. (Credit: Japan Aerospace Exploration Agency)

Japan

Terra

Japan's Ministry of Economy, Trade and Industry (METI) provided the Advanced Spaceborne Thermal Emission and Reflection Radiometer (ASTER) instrument for NASA's Terra satellite. ASTER provides high-spatial-resolution, multispectral images of Earth's surface and clouds. It has 14 spectral bands in the visible-to-thermal infrared wavelength region and high spatial resolution of 15 to 90 meters. Terra was launched in 1999 as the flagship mission of NASA's Earth Observing System. It carries five instruments that observe Earth's atmosphere, oceans, landmasses, snow and ice, and radiant energy. Taken together, these observations provide unique insight into how the Earth system works and how it is changing. Terra had a planned mission life of 6 years and is now operating in extended mission phase.

ASTER is the high-spatial-resolution instrument on the Terra satellite. ASTER's ability to serve as a "zoom" lens for the other Terra instruments is particularly important for change detection, calibration/validation, and land-surface studies. All three ASTER telescopes (visible and near-infrared, short-wavelength infrared, and thermal infrared) are pointable in the cross-track direction. In addition, a second visible and near-infrared camera looks aft along a track and provides the capability to produce high-quality digital terrain topography.

Millions of ASTER scenes and derived products have been distributed to users around the world through the Ground Data System at Japan's Earth Remote Sensing Data Analysis Center and the U.S. Land Processes Distributed Active Archive Center in Sioux Falls, South Dakota. ASTER data and derived products—including detailed maps of land surface temperature, emissivity, reflectance, and elevation—are important scientific tools to users from many Earth science disciplines. The ASTER products are routinely used in studying geologic processes, monitoring land-cover conditions and change, investigating hydrologic resources and processes, monitoring crop condition and development, studying natural disasters (e.g., flooding and volcanic activity), and performing many other important research and practical applications.

For more information on ASTER or NASA's Terra mission, please visit *http://terra.nasa.gov/*, *http://asterweb.jpl.nasa.gov/*, and *http://www.ersdac.or.jp/eng/index.E.html*.

Terra is in a circular Sun-synchronous polar orbit that takes it from north to south (on the daylight side of the Earth) every 99 minutes. Terra collects data about Earth's biogeochemical and energy systems using five sensors that observe the atmosphere, land surface, oceans, snow and ice, and energy budget.

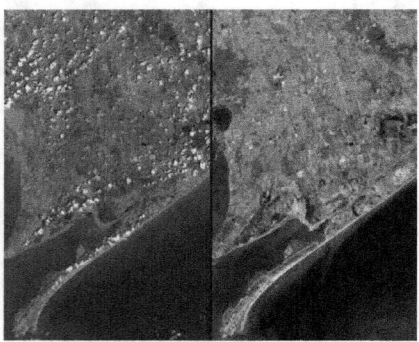

Three weeks after Hurricane Ike came ashore near Galveston, Texas, residents returned to find their houses in ruins. The image on the left was acquired by Terra's ASTER instrument prior to the hurricane. In the image on the right, taken soon after the hurricane, vegetation is displayed in red and inundated areas are shown in blue-green. (Credit: NASA/Goddard Space Flight Center/Japan's Ministry of Economy, Trade, and Industry/Japan Space Systems and U.S.-Japan ASTER Science Team)

Japan

TRMM
Tropical Rainfall Measuring Mission

The Tropical Rainfall Measuring Mission (TRMM), launched by the Japan Aerospace Exploration Agency (JAXA) in 1997, monitors tropical rainfall and the associated release of energy that helps to power the global atmospheric circulation that shapes both weather and climate. NASA provided the spacecraft and four instruments and is responsible for mission operations. In addition to providing the launch, JAXA provided the precipitation radar. TRMM had a planned mission life of 3 years, but is now operating in extended mission phase.

TRMM contributes to the understanding of the role of precipitation in the climate system and how much energy is transported in the global water cycle. TRMM scientists use data from TRMM and other NASA satellites to study the interactions among water vapor, clouds, and precipitation, as well as their role in regulating climate and weather.

The TRMM mission has produced a long-term, high-quality precipitation record and continues to provide real-time data to operational agencies worldwide. Among its many accomplishments, the TRMM mission has: helped to narrow considerably the range of uncertainty in previous space-based rainfall estimates; made possible the quantification of the diurnal cycle of precipitation and convective intensity over land and ocean tropics-wide on fine scales; provided the first comprehensive estimates of how rainfall is directly related to latent heat release in the atmosphere, a key characteristic in understanding the impact of tropical rainfall on the general circulation of the atmosphere; yielded new insights into the dynamics of tropical waves and oscillations, providing the basis of hypotheses on the dynamics of convective-climate feedbacks; and generated data that are used to assess hurricane and typhoon locations and intensities and that have played an important role in the research and analysis of tropical cyclones.

For more information on TRMM, please visit *http://trmm.gsfc.nasa.gov/*.

The TRMM spacecraft obtained this view of Hurricane Irene making landfall along the North Carolina coast in August 2011. The green regions represent areas where the storm is dumping at least half an inch of rain per hour; dark red represents about two inches per hour.

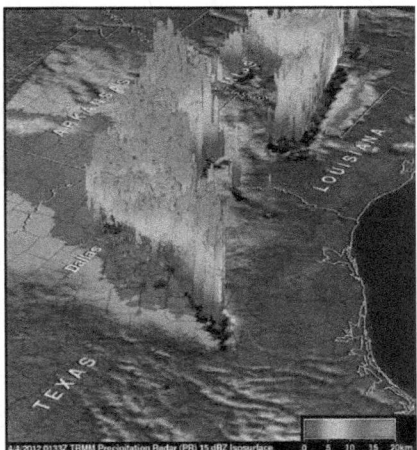

The TRMM satellite passed above tornado-producing storms just east of Dallas, Texas, in April 2012. The TRMM Microwave Imager (TMI) and precipitation radar show a distinct line of tornadic thunderstorms extending from Arkansas through central Texas. The radar data provide a three-dimensional view of the intensity and vertical distribution of precipitation.

Science

NASA astronaut Kevin Ford, International Space Station Expedition 34's commander, is shown in the Destiny Laboratory Module with the Fluids Integration Rack/Light Microscopy Module/Advanced Colloids Experiment.

Republic of Korea

ACE
Advanced Colloids Experiment

The Advanced Colloids Experiment (ACE) is the first in a series of microscopic imaging investigations of materials that contain small colloidal particles, which have the specific characteristic of remaining evenly dispersed and distributed within the material. This investigation takes advantage of the unique environment on board the International Space Station (ISS) to study the effects induced by Earth's gravity on these colloidal materials in order to examine flow characteristics and the evolution and ordering effects within them. Engineering, manipulation, and the fundamental understanding of materials of this nature potentially enhance our ability to produce, store, and manipulate materials that rely on similar physical properties. The ACE investigation consists of 20 sample discs that each contain up to 10 wells of colloidal particles. ACE is a joint experiment between NASA and Chungnam National University (CNU) in South Korea.

The intent of the experiment is to observe, understand, and learn how to control colloidal processes. ACE addresses basic physics questions about colloids, but some of the results may eventually have applications for space exploration. Super-critical fluids, which represent one of the applications of the critical point (i.e., phase boundary) experiment, are a potential application in propulsion systems for future spacecraft design. In addition, associated shelf-life studies impact not only products on store shelves, but also those being stored for later use in space. The ACE samples provide important data that is not available on Earth, such as data that can guide our understanding of crystallization, production quality control, and phase separation (e.g., shelf-life and product collapse). A better understanding of these processes could have an enormous commercial impact in terms of quality, production, and longevity.

The ACE experiment is conducted on board the ISS in the Fluids Integrated Rack (FIR) using the Light Microscopy Module (LMM). NASA is providing four sets of sample modules that will be filled with CNU's sample material. CNU cooperates with NASA on the science investigations and shares the resulting CNU ISS test data. The first 9-day sample run took place in August 2012. The current plan is to launch a set of sample modules about once a year from 2012 through at least 2015, and possibly as late as 2016.

For more information on ACE, please visit *http://www.nasa.gov/mission_pages/station/research/experiments/ACE-1.html#results*.

Republic of Korea

Heliophysics and Space Physics

NASA and the Korea Astronomy and Space Science Institute (KASI) are collaborating in the area of solar and space physics (heliophysics) and space weather research. Heliophysics research and exploration focuses on studying the Sun, the heliosphere, and planetary environments as elements of a single, interconnected system that contains dynamic space weather and evolves in response to solar, planetary, and interstellar conditions. By sharing the data of NASA heliophysics missions—in particular, the Solar Dynamics Observatory (SDO), Van Allen Probes, and Magnetospheric MultiScale (MMS) missions—NASA and KASI will increase their scientific output and productivity to the benefit of heliophysics science and space weather research.

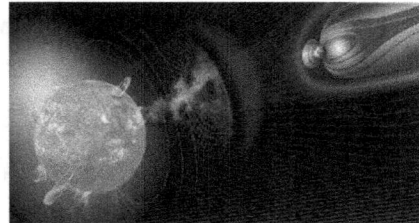

Heliophysics is the study of the Sun and its interactions with Earth and the solar system.

Launched in 2010, SDO is designed to help us understand the Sun's influence on Earth and near-Earth space by studying the solar atmosphere on small scales of space and time and in many wavelengths simultaneously. NASA's Van Allen Probes mission, launched in 2012, provides insight into the dynamics of particle acceleration within the radiation belts and gives scientists the data they need to make predictions of changes in this critical region of space. The MMS mission, scheduled for launch in 2014, will use Earth's magnetosphere as a laboratory to study the microphysics of magnetic reconnection, a fundamental plasma-physical process that converts magnetic energy into heat and the kinetic energy of charged particles.

As part of its agreement with NASA, KASI is building the Korean Data Center for SDO, which will provide free and unfettered access for scientists to the resident archive of SDO data without use and disclosure restrictions. The Korean Data Center for SDO provides scientific data not only to Korean institutes and universities but also to the international space weather community. In addition, in May 2012, KASI dedicated a new, 7-meter antenna that will provide worldwide access to the Van Allen Probes' real-time space weather data.

For more information on NASA and KASI collaboration, please visit *http://www.nasa.gov/topics/solarsystem/sunearthsystem/main/News081210-kasi.html*.

For more information on the NASA Heliophysics Research Program, please visit *http://science.nasa.gov/heliophysics/*.

Science

An artist's concept of Aura satellite orbiting Earth.

Tropospheric NO_2 Column Density (x 10^{15} molecules/cm^2)

0.0   0.1 0.2   0.5   1.0   2.0

Data from the Ozone Monitoring Instrument on NASA's Aura satellite show long, faint tracks of elevated nitrogen dioxide (NO_2) levels along certain shipping routes. For more than a decade, scientists have observed "ship tracks" in natural-color satellite imagery of cloud layers over the ocean. These bright trails are a visible sign of pollution from ship exhaust, and scientists can now see that ships have a more subtle signature as well.

The Netherlands

Aura

NASA's Aura (Latin for "breeze") satellite, launched by NASA in 2004, measures changes in the composition, chemistry, and dynamics of the Earth's atmosphere for research and applications that are relevant to ozone trends, air quality, and climate. Aura is operating in a Sun-synchronous, near-polar, low-Earth orbit. The Netherlands, in cooperation with the Finnish Meteorological Institute (FMI), provided Aura's Ozone Monitoring Instrument (OMI).

OMI is a nadir-viewing wide-field imaging spectrometer that measures the solar radiation backscattered by the Earth's atmosphere and surface with a wide-field telescope feeding two imaging grating spectrometers. Each spectrometer employs a charge-couple device (CCD) detector. OMI measures the complete spectrum in the ultraviolet/visible wavelength range with a very high spectral resolution. Daily global coverage is achieved with a spatial resolution of about 15 by 25 kilometers.

OMI measures total ozone and other atmospheric parameters related to ozone chemistry and climate. OMI can distinguish between aerosol types (i.e., smoke, dust, and sulfates) and can measure both the cloud pressure and coverage, thereby providing the data to derive tropospheric ozone. OMI maps pollution products on urban scales and tracks its transcontinental transport.

The OMI International Science Team consists of scientists from the Netherlands, Finland, and the United States. It is responsible for algorithm development, in-flight instrument calibration and trend monitoring, data processing, validation, and analysis.

Aura has provided valuable ozone measurements; tropospheric maps of carbon monoxide, water vapor, and cloud ice; and measurements of the stratosphere. These measurements enable scientists to study ozone trends, air quality changes, and their linkage to climate change.

For more information on the OMI and Aura, please visit *http://aura.gsfc.nasa.gov/index.html*, *http://aura.gsfc.nasa.gov/instruments/omi/index.html*, and *http://www.knmi.nl/omi/publ-en/news/index_en.html*.

The Netherlands

Flight Deck System Design

A flight deck is the volume of space on an aircraft that is designed to accommodate at least one human operator and to provide interfaces between the operator(s) and the remainder of the flight deck system. The flight deck system includes the entity or entities that have the authority and responsibility for directing the flight of an aircraft, all subsystems that directly interface to the entity or entities, and all interfaces between them.

NASA and the Netherlands' National Aerospace Laboratory (NLR) are pursuing collaborative research in state-of-the-art flight-deck-system design methods and tools. Experiments will be conducted that are related to shared situation awareness, command generation, action coordination, conflict detection and resolution, and the management of automation-initiated behaviors such as hazard avoidance and self-separation from aircraft. The overarching objective of the experiments will be to evaluate design concepts against their safety potential and to validate modeled predictions of performance and safety attributes.

NASA believes that future flight deck systems should systematically incorporate integrated displays, decision-making functions, information management, and dynamically allocated human/automation task responsibilities. The ideal future flight deck system would be aware of the state of the vehicle and the operator and would respond appropriately. The system would sense internal and external hazards, evaluate them, and provide key information to facilitate timely and appropriate responses. The system would also be robust and adaptable to the addition of new functions and information sources as they become available.

For more information on NLR, please visit *http://www.nlr.nl/*.

The flight deck of the NASA B-737 Flying Laboratory.

The NASA B-737 Flying Laboratory.

Science

Norway

IRIS
Interface Region Imaging Spectrograph

Understanding the interface between the photosphere and corona remains a fundamental challenge in solar and heliospheric science. The Interface Region Imaging Spectrograph (IRIS) mission opens a window of discovery into this crucial region by tracing the flow of energy and plasma through the chromosphere and the transition region into the corona using spectrometry and imaging. Launched by NASA in 2013, IRIS is NASA-led mission with important support from the Norwegian Space Centre (NSC). The primary goal of IRIS is to better understand how the solar atmosphere is energized.

IRIS is designed to provide significant new information to increase our understanding of energy transport into the corona and solar wind and to provide an archetype for all stellar atmospheres. IRIS's unique instrument capabilities, coupled with state-of-the-art three-dimensional (3-D) modeling, will fill a large gap in our knowledge of this dynamic region of the solar atmosphere. The mission will extend the scientific output of existing heliophysics spacecraft that follow the effects of energy release processes from the Sun to Earth. By combining high-resolution imaging and spectroscopy for the entire chromosphere and adjacent regions, IRIS will advance our understanding of the solar drivers of space weather from the corona to the far heliosphere. IRIS will resolve in space, time, and wavelength the dynamic geometry from the chromosphere to the low-temperature corona to shed much-needed light on the physics of this magnetic interface region.

NASA and NSC will collaborate on IRIS observations, which will be collected through the ground station located at the NSC Kongsberg Satellite Services station in Svalbard, Norway. NASA and NSC will work together to explore the solar chromosphere and transition region through the use a solar telescope and spectrograph. NSC will also support joint scientific analysis activities in Norway and the United States, including analysis of the IRIS observations using 3-D numerical models from the Institute of Theoretical Astrophysics at the University of Oslo, Norway.

For more information on IRIS, please visit *http://iris.lmsal.com/*.

An artist's depiction of the IRIS spacecraft.

The fully integrated spacecraft and science instrument for NASA's IRIS mission is seen in a clean room at the Lockheed Martin Space Systems facility in Sunnyvale, California. The solar arrays are deployed in the configuration they will assume when in orbit. (Credit: Lockheed Martin)

Norway

Ka-band Radio Frequency (RF) Propagation Monitoring Station

NASA and the Norwegian Space Centre (NSC) are cooperating on an experiment to determine the expected Ka-band performance of the forthcoming upgrades to the Near Earth Network (NEN) ground station operating at Svalbard, Norway, with respect to atmospheric propagation effects. In particular, validation of anticipated atmospheric-induced attenuation losses and increases in system temperature are being conducted by deploying a microwave radiometer specifically designed to measure how much the atmosphere contributes to Ka-band link degradation.

Conducting this measurement campaign will directly provide enhanced system planning through the accurate determination of expected link availability at Ka-band, improved science mission planning by managing expectations and maximizing mission success and data throughput, and necessary preparation for deployment of the NEN Ka-band Polar Network.

These studies are being carried out in preparation for supporting high-priority Ka-band NASA Earth Science missions such as Surface Water Ocean Topography (SWOT). The propagation data gathered by the monitoring station will be used to optimize key specifications for the Ka-band link.

Data collection began in early 2011 and will continue through 2015.

For more information, please visit *http://ntrs.nasa.gov/archive/nasa/casi.ntrs.nasa.gov/20120000737_2012000660.pdf*.

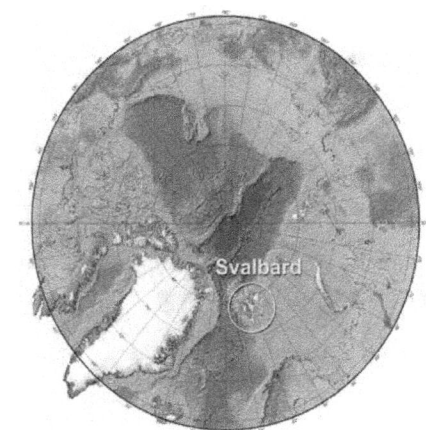

Svalbard is located in the northernmost part of Norway, within the Arctic Circle.

Human Exploration and Operations

Norway

NASA Isbjørn Facility at SvalSat

The Norwegian Space Centre (NSC) provided the facility infrastructure at the Svalbard Satellite Station (SvalSat) in Longyearbyen, Svalbard, to support the establishment and operation of the NASA Isbjørn Facility, a satellite- and launch-vehicle-tracking, data-acquisition, and control facility. The NASA Isbjørn Facility and SvalSat have been in operation since 1997.

SvalSat is the northernmost ground station in the world, and it provides telemetry, tracking, and command (TT&C) and data reception for polar-orbiting satellites. It also provides launch and early orbit support for orbital launches, as well as tracking of sounding rockets.

The NSC provided the infrastructure at SvalSat for the NASA Isbjørn Facility, including installation of basic utilities and communications, such as power, water, and telephone; installation of a fiber optic telecommunications link from Svalsat to Longyearbyen; construction of a station building to house SvalSat tenants, including NASA and the U.S. National Oceanic and Atmospheric Administration (NOAA); coordination of SvalSat activity with Norwegian Government officials and regulatory bodies; and spectrum management.

NASA provided, installed, and currently contracts for the operation and maintenance by Kongsberg Satellite Services (KSAT) of the NASA Isbjørn Facility, including all of the instrumentation, one 11-meter S-X band antenna, and associated ground support equipment. NASA uses the Isbjørn Facility to provide tracking and data-acquisition services for polar-orbiting Earth observation satellites, including Landsat-7.

In 2003, NASA and NOAA provided for the installation by KSAT of approximately 3,000 kilometers of fiber optic cable from SvalSat to the Andøya Rocket Range in Norway, and from Andøya to New York. This fiber optic communication network replaced the need for satellite communications services and enables real-time transmission and processing of tracking, data acquisition, and control data for NASA's Earth science satellites at a substantial savings to NASA.

SvalSat currently has six antenna systems, including two operational 11-meter systems and one 13-meter S- and X-band system. NASA has access to three of these antennas under a pooled-service concept.

For more information on the NASA Isbjørn Facility at SvalSat, please visit *http://scp.gsfc.nasa.gov/gn/norway.htm*.

Four of the six protective radar domes (radomes) over the telemetry and command antennas at SvalSat.

A close-up of one of the six radomes at SvalSat.

Russia

ISS
International Space Station

The Russian Federal Space Agency's (Roscosmos) primary contributions to the International Space Station (ISS) Program are Soyuz crew transportation, Progress logistics transportation, the Zvezda Service Module, the Pirs Airlock and Docking Compartment, the Multipurpose Laboratory Module (MLM), and the U.S.-owned/Russian-built Zarya Functional Cargo Block (FGB).

The ISS Program is a partnership among the space agencies of Canada, Europe, Japan, Russia, and the United States. International crews have continuously inhabited the ISS since November 2000. Cargo transportation is provided by European, Russian, and Japanese vehicles and by NASA's contracts for Commercial Resupply Services (CRS). Crew transportation and rescue is provided by the Russian Federal Space Agency and, beginning in 2017, by NASA's commercial crew transportation services.

NASA's contributions to the ISS Program include overall management, coordination, and integration of the ISS Program across the partnership for operations, safety, transportation, and integrated hardware performance. NASA hardware components include the U.S. laboratory module, the airlock for extravehicular activities (EVA), truss segments to support the U.S. solar arrays and thermal radiators, three connecting nodes, and living quarters.

The ISS functions as an orbital microgravity and life sciences laboratory, a test bed for new technologies in areas such as life support and robotics, and a platform for astronomical and Earth observations. The ISS also serves as a unique engineering test bed for flight systems and operations critical to NASA's future exploration missions. U.S. research on the ISS concentrates on the long-term effects of space travel on humans and on technology development activities in support of exploration. The ISS is a stepping-stone for human exploration and scientific discovery beyond low-Earth orbit.

Zarya, the first component of the ISS, was built in Russia under a U.S. contract and, as such, is a U.S. contribution to the ISS. Zvezda is the structural and functional center of the Russian segment of the ISS. The Pirs Airlock provides the capability for EVAs from Zvezda and the Docking Compartment provides a port for docking Soyuz and Progress logistics vehicles. The MLM will be used for experiments, docking, and cargo, and it will serve as a crew work and rest area. The Soyuz vehicle transports three crewmembers to the ISS and serves as the emergency lifeboat should the crew need to return to Earth. The Progress is a resupply vehicle used for cargo and propellant deliveries. Progress engines can boost the ISS to higher altitudes and perform orientation control of the ISS.

For more information on the ISS and Russia's contributions, please visit *http://www.nasa.gov/mission_pages/station/structure/assembly_elements.html*, and *http://www.federalspace.ru/?lang=en*.

Soyuz TMA-7 orbiting Earth. The Soyuz is a crew transportation and rescue vehicle and can support up to three crewmembers.

This photo, taken by astronauts on board the International Space Station (ISS), shows the Russian Soyuz spacecraft (left) and the Progress spacecraft (right) docked to the ISS as it flies over the Nile River Delta in Egypt. The cities of Cairo and Alexandria are both visible in the frame. (Credit: NASA, European Space Agency)

Science

Russia

LRO
Lunar Reconnaissance Orbiter

An artist's rendering of the Lunar Reconnaissance Orbiter leaving Earth orbit for its journey to the Moon.

The Russian Federal Space Agency (Roscosmos) provided the Lunar Exploration Neutron Detector (LEND) instrument for NASA's Lunar Reconnaissance Orbiter (LRO), which NASA launched in June 2009.

The primary objective of the LRO is to obtain data that will facilitate returning humans safely to the Moon and enable extended stays. The high-priority investigations defined for LRO included high-resolution imaging for characterizing landing site selection and safety, geodetic global topography for the next steps in exploration, measurements to globally assess polar resources and their accessibility for human exploration, and radiation measurements for future human exploration on the Moon.

The LEND team is led by the Space Research Institute of the Russian Academy of Sciences (IKI RAS) and consists of scientists from IKI RAS, the University of Arizona, the University of Maryland, and NASA's Goddard Space Flight Center (GSFC). One of six instruments aboard the LRO, LEND has passive collimators of neutrons that provide high-spatial-resolution maps of neutron emission at the lunar surface to search for evidence of water ice and provide measurements of the lunar radiation environment—crucial information for future human exploration.

A picture of the Lunar Exploration Neutron Detector instrument. (Credit: Space Research Institute of the Russian Academy of Sciences).

Using LEND data as a primary source of information along with the results of other LRO instruments, NASA chose Cabeus crater as a location for lunar surface bombardment by NASA's Lunar Crater Observation and Sensing Satellite (LCROSS). This experiment studied the composition of the lunar surface to look for water ice by measuring the parameters of debris plume. The resulting data confirmed the presence of water ice and characterized its distribution in permanently shadowed regions of the moon. The availability of water resources could enable the production of oxygen, drinking water, and rocket fuel—all of which have previously been serious barriers to human exploration.

LEND is a modified and enhanced version of the Russian High-Energy Neutron Detector (HEND), which launched in 2001 aboard NASA's Mars Odyssey spacecraft and continues to look for water deposits on Mars.

For more information on LEND, please visit *http://lro.gsfc.nasa.gov/* and *http://lunar.gsfc.nasa.gov/lend.html*.

Russia

Mars Odyssey

The Russian Federal Space Agency (Roscosmos), along with its participating entity, the Space Research Institute of the Russian Academy of Sciences (IKI RAS), provided the High-Energy Neutron Detector (HEND) instrument for NASA's 2001 Mars Odyssey mission.

Mars Odyssey mapped, for the first time, the amount and distribution of chemical elements and minerals that make up the Martian surface. Odyssey's maps of hydrogen distribution led scientists to discover vast amounts of water ice in the polar regions of Mars, buried just beneath the surface. Odyssey recorded the radiation environment in low-Mars orbit to determine the radiation-related risk to future human explorers who may one day go to Mars.

The HEND is a part of the Gamma Ray Spectrometer (GRS) instrument suite, one of three primary instrument packages on the Mars Odyssey. The GRS instrument suite has provided data on the elemental distribution at the surface of Mars and searched for water across the planet. Specifically, this instrument suite detects and counts gamma rays and neutrons from the Martian surface. By associating the spectral distribution of the gamma rays with known nuclear transitions lines, it is possible to calculate the ratio of elemental abundances on the surface. By counting the number of neutrons as a function of energy, it is possible to calculate the hydrogen abundance, thus inferring the presence of water.

In 2008, the Odyssey orbiter found salt deposits in the southern highlands of Mars, in basins with channels leading into them. These features are reminiscent of salt-pan deposits on Earth, which are consistent with water flowing over a long period of time. The salt deposits could have come from water that had evaporated, leaving mineral deposits that had built up over the years. The presence of these salt deposits could aid in answering the most important scientific question driving Mars research: did life ever exist on the Red Planet?

For more information on the Mars Odyssey, please visit *http://marsprogram.jpl.nasa.gov/odyssey/overview/*.

For more information on the HEND instrument, please visit *http://www.iki.rssi.ru/hend/index-engl.htm*.

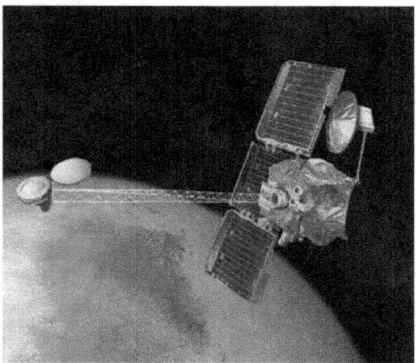

An artist's concept of the 2001 Mars Odyssey orbiting Mars.

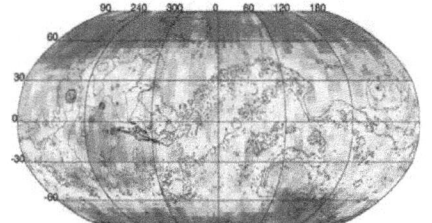

Observations by Russia's High-Energy Neutron Detector on NASA's 2001 Mars Odyssey spacecraft show a global view of Mars in high-energy "fast" neutrons. The low flux of fast neutrons (blue and purple colors) in the north polar region suggests an abundance of hydrogen in the soil. (Credit: NASA/Jet Propulsion Laboratory/University of Arizona/Roscosmos/Space Research Institute)

Russia

MSL
Mars Science Laboratory

This image of NASA's Curiosity rover shows the location of the two components of the Dynamic Albedo of Neutrons instrument. The neutron generator is mounted on the right hip (visible in this view), and the detectors are on the opposite hip. (Credit: NASA/Jet Propulsion Laboratory–Caltech)

NASA's Mars Science Laboratory (MSL) mission utilized an innovative sky crane to set down the Curiosity rover onto the surface of Mars in August 2012, where it began its mission to assess whether Mars has ever had an environment capable of supporting small life forms called microbes. The MSL mission, launched by NASA in November 2011, is planned to last at least one Martian year (687 Earth days), but it has the potential to operate for much longer.

The primary scientific objective, to be carried out during the surface science phase of the MSL mission, is to assess the biological potential of at least one target area by characterizing the local geology and geochemistry, investigating planetary processes relevant to habitability (including the role of water), and characterizing the broad spectrum of surface radiation. The landing site, Gale Crater, was selected based on an assessment of safety and planetary protection and an analysis by the scientific community. Since its landing, MSL has already begun to make new and exciting science discoveries, such as observing a natural intersection of three kinds of terrain at an area called Glenelg and finding evidence that a stream once ran across the area on Mars where the rover is driving. Although earlier missions detected the presence of water on Mars, the images taken by MSL of rocks that contain ancient streambed gravel are the first of their kind.

The Space Research Institute of the Russian Academy of Sciences (IKI RAS) provided the Dynamic Albedo of Neutrons (DAN) detector to the MSL mission. DAN searches for water by looking for neutrons escaping from the planet's surface. Cosmic rays from space constantly bombard the surface of Mars, knocking neutrons in surface soils and rocks out of their atomic orbits. If liquid or frozen water happens to be present, hydrogen atoms in water molecules slow the neutrons down. As a result, some of the neutrons escaping into space have less energy and move more slowly. These slower particles can be measured with a neutron detector such as DAN. Scientists expect to find hydrogen on Mars in two forms: water ice and minerals that have molecules of water in their crystal structures. DAN's pulsing neutron generator is sensitive enough to detect water content as low as one-tenth of one percent and resolve layers of water and ice beneath the surface. DAN will focus a beam of neutrons onto the Martian surface. The neutrons are expected to travel 1–2 meters below the surface before being absorbed by hydrogen atoms in subsurface ice.

For more information on the MSL mission and the DAN instrument, please visit *http://mars.jpl.nasa.gov/msl/* and *http://marsprogram.jpl.nasa.gov/msl/mission/instruments/radiationdetectors/dan/*.

Russia

Wind

NASA's Wind satellite, launched in 1994, became the first of two NASA spacecraft in the Global Geospace Science (GGS) Program and was also part of the International Solar-Terrestrial Physics (ISTP) Science Initiative. The Wind mission includes the Russian Konus experiment, which was the first Russian scientific instrument to fly on an American satellite.

The Wind satellite operates at the sunward Sun-Earth Lagrange point (L1), giving scientists an opportunity to study the enormous flow of energy and momentum known as the solar wind. Wind measures the mass, momentum, and energy of the solar wind that is transferred into the space environment around Earth.

The Konus experiment provides omnidirectional and continuous coverage of cosmic gamma-ray transients. The instrument monitors cosmic gamma-ray bursts, soft gamma repeaters (SGR), solar flares, and other transients with the moderate energy resolution available from scintillation spectrometers. In conjunction with other instruments, Konus has helped to determine gamma-ray burst locations, thereby enabling the prompt and ongoing study of this elusive phenomenon. NASA's Wind satellite remains in operation.

For more information on Wind/Konus, please visit *http://pwg.gsfc.nasa.gov/wind.shtml*.

An artist's concept of the Wind spacecraft.

Human Exploration and Operations

Spain

DSN
Deep Space Network

The Spanish National Institute of Aerospace Technology (INTA), on behalf of NASA, manages the Madrid Deep Space Communications Complex (MDSCC), located 60 kilometers west of Madrid, Spain, in Robledo de Chavela. The MDSCC has been in operation since 1964 as one of three facilities that comprise NASA's Deep Space Network (DSN).

NASA's DSN is an international network of antennas that communicates with interplanetary spacecraft, is used by radio and radar astronomers to observe the solar system and the universe, and supports selected Earth-orbiting satellites. NASA's three DSN facilities are Goldstone, in the Mojave Desert, California; the Canberra Deep Space Communication Complex (CDSCC), near Canberra, Australia; and MDSCC, near Madrid, Spain.

These DSN facilities are located approximately 120° apart around the world, which permits constant observation of spacecraft as the Earth rotates. Each location has an 8- to 14-hour viewing period for contact with spacecraft. The DSN provides the two-way communications link that tracks, guides, and controls the spacecraft and returns the telemetry and scientific data collected by the spacecraft. Incoming data to the three DSN facilities are processed and transmitted to NASA's Jet Propulsion Laboratory (JPL) in Pasadena, California, for further processing and distribution to science teams over a modern ground-communications network.

Each DSN facility consists of three classes of deep space antennas: a 34-meter-diameter high-efficiency antenna, a 34-meter beam waveguide antenna, and a 70-meter antenna.

For more information on the DSN and MDSCC, please visit *http://deepspace.jpl.nasa.gov/dsn/* and *http://www.mdscc.org/*.

Deep Space Station–63 (DSS-63) at Madrid Deep Space Network Complex.

An aerial view of Madrid Deep Space Network Complex.

Spain

MSL
Mars Science Laboratory

NASA's Mars Science Laboratory (MSL) mission utilized an innovative sky crane to set down the Curiosity rover onto the surface of Mars in August 2012, where it began its mission to assess whether Mars has ever had an environment capable of supporting small life forms called microbes. The MSL mission, launched by NASA in November 2011, is planned to last at least one Martian year (687 Earth days), but has the potential to operate for much longer.

The primary scientific objective, to be carried out during the surface science phase of the MSL mission, is to assess the biological potential of at least one target area by characterizing the local geology and geochemistry, investigating planetary processes relevant to habitability (including the role of water), and characterizing the broad spectrum of surface radiation. The landing site, Gale Crater, was selected based on an assessment of safety and planetary protection and an analysis by the scientific community. Since its landing, MSL has already begun to make new and exciting science discoveries, such as observing a natural intersection of three kinds of terrain at an area called Glenelg and finding evidence that a stream once ran across the area on Mars where the rover is driving. Although earlier missions detected the presence of water on Mars, the images taken by MSL of rocks that contain ancient streambed gravel are the first of their kind.

The Spanish Center for Astrobiology (CAB), part of the National Institute for Aerospace Technology (INTA), provided MSL's Rover Environmental Monitoring Station (REMS) instrument suite. Spain's Center for the Development of Industrial Technology (CDTI) also provided components of the rover high-gain antenna subsystem. REMS provides a daily report of atmospheric weather conditions on Mars. From its vantage point attached to the vertical mast on the MSL deck, REMS measures atmospheric pressure, humidity, ultraviolet radiation from the Sun, wind speed, wind direction, ground temperature, and air temperature.

For more information on the MSL mission and the REMS instrument, please visit *http://mars.jpl.nasa.gov/msl/* and *http://marsprogram.jpl.nasa.gov/msl/mission/instruments/environsensors/rems/*.

This image from NASA's Curiosity rover shows the first sample of powdered rock extracted by the rover's drill and transferred from the drill to the rover's scoop. The scoop is 1.8 inches (4.5 centimeters) wide. The image was obtained by Curiosity's Mast Camera in February 2013. (Credit: NASA/Jet Propulsion Laboratory–Caltech/Malin Space Science Systems)

Science

An artist's concept of the Fermi spacecraft orbiting Earth.

Sweden

Fermi Gamma-ray Space Telescope

The Fermi Gamma-ray Space Telescope, launched by NASA in 2008, is a spacecraft with the ability to detect gamma rays created by the most energetic objects and phenomena in the universe. Among the topics of cosmological interest studied by the mission are dark matter and the periods of star and galaxy formation in the early universe. The mission is an astrophysics and particle physics partnership developed by NASA in collaboration with the U.S. Department of Energy, with important contributions from academic institutions and partners in France, Germany, Italy, Japan, Sweden, and the United States.

Because of their tremendous energy, gamma rays travel through the universe largely unobstructed. This means that Fermi is able to observe gamma-ray sources located near the edge of the visible universe. Gamma rays detected by Fermi originate near the otherwise-obscured central regions of exotic objects like supermassive black holes, pulsars, and gamma-ray bursts.

In order to study these high-energy waves, two main instruments are used: the Large Area Telescope (LAT) and the Gamma-ray Large Area Space Telescope (GLAST) Burst Monitor (GBM). When in survey mode, the LAT scans the entire sky every 3 hours. The instrument detects photons with energies ranging from 20 million electron volts to over 300 billion electron volts. In its first month of operation, the GBM instrument spotted 31 gamma-ray bursts, which occur when massive stars die or when orbiting neutron stars spiral together and merge. By the end of 2011, Fermi had detected more than 100 gamma-emitting pulsars; before its launch, only 7 of these were known to emit gamma rays. Newly developed methods of analysis promise to spur additional discoveries.

The Swedish Royal Institute of Technology (KTH) provided sensors for the LAT Calorimeter. Along with Sweden and NASA, the following organizations are also contributors to the Fermi mission: the German Aerospace Center (DLR), Italian Space Agency (ASI), Japan Aerospace Exploration Agency (JAXA), and French Alternative Energies and Atomic Energy Commission (CEA).

For more information on the FERMI mission, please visit *http://fermi.gsfc.nasa.gov/*.

Sweden

MMS
Magnetospheric MultiScale

NASA's Magnetospheric MultiScale (MMS) mission is a solar-terrestrial probe mission comprising four identically instrumented spacecraft that will use Earth's magnetosphere as a laboratory to study the microphysics of three fundamental plasma processes: magnetic reconnection, energetic particle acceleration, and turbulence. These processes occur in all astrophysical plasma systems but can be studied in situ only in our solar system and most efficiently only in Earth's magnetosphere, where they control the dynamics of the geospace environment and play an important role in the processes known as "space weather." The four MMS spacecraft are currently planned for launch by NASA in 2014.

NASA is collaborating on the MMS mission with the Swedish National Space Board (SNSB), with support from the Royal Institute of Technology (KTH) and the Swedish Institute of Space Physics (IRF). NASA is also cooperating with the Aeronautics and Space Agency (ALR) of the Austrian Research Promotion Agency (FFG), the Japan Aerospace Exploration Agency (JAXA), and the French National Center for Space Studies (CNES) on the mission.

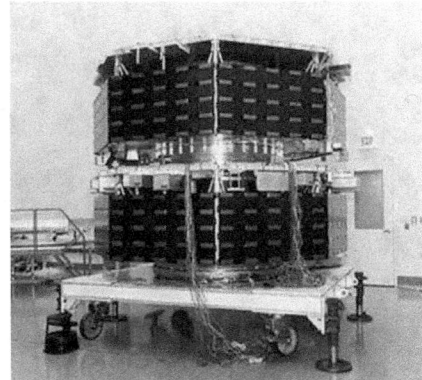

Flight structures for MMS #3 (top) and MMS #4 (bottom) are shown after being integrated into a "mini-stack" testing configuration for environmental testing at NASA's Goddard Space Flight Center. This is also the configuration that will be used for shipping the four spacecraft to Cape Canaveral, Florida, for launch. The structures seen here include the flight thrust tubes, spacecraft, and Instrument decks, in addition to mass simulators for the solar arrays and instruments.

MMS investigates how the magnetic fields of the Sun and Earth connect and disconnect, explosively transferring energy from one to the other in a process known as magnetic reconnection. This process limits the performance of fusion reactors and is the final governor of geospace weather, which affects modern technological systems such as telecommunications networks, Global Positioning System (GPS) navigation, and electrical power grids. MMS's four spacecraft will measure plasmas, fields, and particles in a near-equatorial orbit that will frequently encounter reconnection in action.

The four MMS spacecraft will carry identical instrument suites of plasma analyzers, energetic particle detectors, magnetometers, and electric field instruments; they will also carry a device to prevent spacecraft charging from interfering with the highly sensitive measurements required in and around the diffusion regions. SNSB, through KTH and IRFU, is designing and developing the Spin-Plane Double Probe (SDP) instrument for measuring the direct current electric fields on the MMS spacecraft, as well as the Low-Voltage Power Supply (LVPS) for the MMS FIELDS investigation for the four MMS spacecraft. SNSB will also support the integration and testing of the SDP and LVPS for the instrument suite. NASA is providing the spacecraft, the launch, and overall mission management.

For more information on MMS, please visit *http://mms.gsfc.nasa.gov/*.

Science

Switzerland

IBEX
Interstellar Boundary Explorer

NASA's Interstellar Boundary Explorer (IBEX), launched in 2008, is the first spacecraft designed to collect data across the entire sky about the heliosphere and its boundary. Scientists have used this data to construct the first comprehensive map of the boundary of our solar system and its location in the Milky Way galaxy. IBEX, which is orbiting Earth just slightly outside of its magnetosphere, is discovering what the interactions are between the solar wind caused by the Sun and the interstellar medium that exists at the edge of our solar system. The University of Bern in Switzerland is cooperating with NASA on this mission.

Scientific results from the IBEX spacecraft have shown that our solar system is traveling slower with respect to the interstellar medium, and that interstellar material is coming into our solar system from a slightly different direction than previously thought. IBEX has also captured the best and most complete glimpse yet of what lies beyond the solar system. These new measurements give clues about how and where our solar system formed, the forces that physically shape our solar system, and the history of other stars in the Milky Way. IBEX has also been able to create new maps that help scientists understand parts of the universe that were once unseen, to place these observations in the context of the Sun's variability, and to describe how that variability interacts with the interstellar medium just beyond the heliosphere. One of the most exciting discoveries of the mission was the detection of an unsuspected ribbon of energetic neutral particles located just outside our solar system.

The University of Bern provided the IBEX-Hi Pre-collimator, the IBEX-Lo Pre-collimator, and the IBEX-Lo Outer Electrostatic Analyzer for the mission. IBEX also features a pair of energetic neutral atom (ENA) cameras to observe the global interactions between the solar wind and interstellar medium, providing a much deeper understanding of the Sun's interaction with the galaxy. From this data, maps of the boundary are created.

For more information on IBEX, please visit *http://www.ibex.swri.edu/*.

An artist's concept of the IBEX spacecraft. (Credit: NASA/Goddard Conceptual Image Lab)

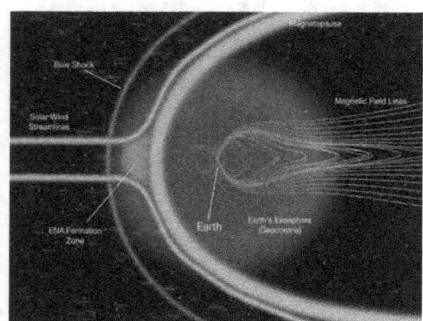

An artist's representation of Earth's magnetosphere; the pink area in front of Earth's magnetopause is the region where IBEX has detected energetic neutral atoms. Our magnetosphere boundary does not give off light that we can detect, so we must use particle sensors like those of IBEX to study regions like this.

Thailand

Vector-Borne Disease Control Research

Mosquitoes are estimated to transmit disease to more than 700 million people annually, resulting in over a million deaths each year. NASA and Mahidol University have been collaborating since 2001 to develop new techniques using remote sensing technology to assist in the monitoring and control of vector-borne diseases. The scientific objectives of the joint research include mapping the breeding sites for the major vector species, identifying potential sites for larvicide and insecticide applications, exploring the linkages of vector population and transmission intensity to environmental variables, monitoring the impact of climate change and human activities on vector population and transmission, and developing a predictive model for disease distribution.

This activity is part of NASA's Applied Sciences program area, which focuses on Earth science applications to public health, particularly regarding infectious disease, emergency preparedness and response, and environmental health issues. The application focus area explores issues of toxic and pathogenic exposure, as well as natural and humanmade hazards and their effects, for risk characterization/mitigation and improvements to health and safety.

For more information on Vector-Borne Disease Control Research, please visit *http://www.sc.mahidol.ac.th/research/vectors.html*.

Thailand's Chao Phraya River forms at the confluence of smaller rivers near Nakhon Sawan and flows past Bangkok to the Gulf of Thailand. In October 2011, Thailand experienced massive flooding, including in Bangkok. As rivers overflowed in Thailand, the Tônlé Sab and the Mekong River remained swollen in neighboring Cambodia. The Moderate Resolution Imaging Spectroradiometer (MODIS) on NASA's Terra satellite captured the top image in October 2011.

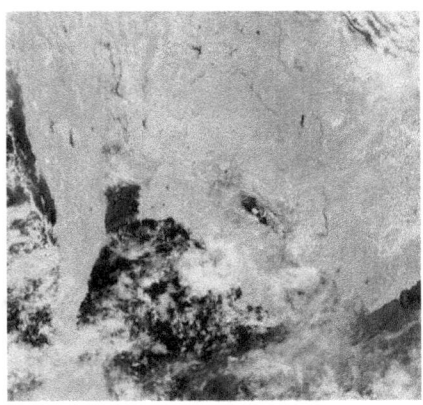

For comparison, the bottom image shows the same area in October 2010. These images use a combination of visible and infrared light to better distinguish between water and land. Vegetation is green, clouds are pale blue-green, and water is dark blue. In 2011, water rests on floodplains between Phitsanulok and Nakhon Sawan. Meanwhile, Tônlé Sab has spread well beyond its 2010 boundaries. The presence of stagnant water, and the fact that many residents were forced to live in close quarters with livestock, contributed to the toll from vector-borne diseases such as Dengue fever.

Science

An artist's concept of the Aura satellite orbiting Earth.

United Kingdom

Aura

NASA's Aura (Latin for "breeze") satellite, launched by NASA in 2004, measures changes in the composition, chemistry, and dynamics of the Earth's atmosphere for research and applications that are relevant to ozone trends, air quality, and climate. Aura is operating in a Sun-synchronous, near-polar low-Earth orbit.

The National Environment Research Council (NERC) of the United Kingdom, in cooperation with the United Kingdom Space Agency, is collaborating with NASA on the jointly developed High-Resolution Dynamic Limb Sounder (HIRDLS) instrument.

HIRDLS is one of four instruments on board Aura. The other three instruments are the NASA-provided Microwave Limb Sounder (MLS), the NASA-provided Tropospheric Emission Spectrometer (TES), and the Netherlands-provided Ozone Monitoring Instrument (OMI). These instruments offer unique and complementary capabilities that enable daily global observations of Earth's atmospheric ozone layer, air quality, and key climate parameters contributing to Aura's overall mission.

HIRDLS is a 21-channel, infrared, limb-scanning radiometer designed to obtain profiles over most of the globe, both day and night, and to provide complete Earth coverage every 12 hours. HIRDLS is designed to sound the upper troposphere, stratosphere, and mesosphere in order to determine temperature and the concentrations of gases and aerosols crucial to ozone chemistry and climate. The locations of polar stratospheric clouds and cloud tops are particularly important. To improve our understanding of atmospheric processes, HIRDLS provides vertical concentrations throughout the stratosphere with improved sensitivity, accuracy, and vertical resolution.

The Aura instruments have provided valuable ozone measurements; tropospheric maps of carbon monoxide, water vapor, and cloud ice; and measurements of the stratosphere. These measurements enable the investigation of questions about ozone trends, air quality changes, and their linkage to climate change.

For more information on the HIRDLS, please visit *http://aura.gsfc.nasa.gov/index.html* and *http://aura.gsfc.nasa.gov/instruments/hirdls/index.html*.

United Kingdom

Hinode

Hinode is a Japan Aerospace Exploration Agency (JAXA)–led solar physics mission that is providing the first solar optical telescope in space. NASA and the United Kingdom (U.K.) are collaborators on the mission.

An artist's concept of the Hinode spacecraft.

Launched by JAXA in 2006 from the Uchinoura Space Center in Japan, Hinode data is helping scientists better understand how magnetic fields interact with plasma to produce solar variability and how the solar photosphere and corona act as a system. The Hinode mission aims to solve how the generation of the solar-magnetic field and its emergence through the photosphere governs the structure of the entire solar atmosphere.

The Hinode mission includes a suite of three science instruments—the Solar Optical Telescope (SOT), the x-ray telescope (XRT), and the Extreme-ultraviolet Imaging Spectrometer (EIS)—to study the interaction between the Sun's magnetic field and its high-temperature, ionized atmosphere. Hinode is circling Earth in a polar, Sun-synchronous orbit that allows the spacecraft's instruments to remain in continuous sunlight for 9 months each year.

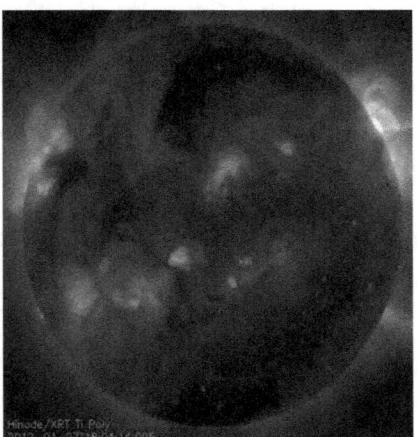
This x-ray image of the Sun was taken by the Hinode X-ray Telescope.

Hinode's instruments have produced fantastic detail of both visible and magnetic features on the Sun's surface and in its atmosphere, the corona. Since 2006, the Hinode mission has helped explain the origin of the solar wind, discovered potential candidates for how the corona gets so hot, and provided images of the complex magnetic structures looping up and out of active regions on the Sun. Hinode's Solar Optical Telescope has delivered images that show greatly magnified views of the Sun's surface. These images are revealing new details about solar convection, the process that drives the rising and falling of gases in the lowest atmospheric region, the photosphere. Hinode's Solar Optical Telescope is the first space-borne instrument to measure the strength and direction of the Sun's magnetic field.

The EIS was built by a consortium led by the U.K.'s Mullard Space Science Laboratory. The U.K. also provided overall program management and system engineering functions for the spectrometer design, development, integration, and test activities, in addition to scientific support for all phases of the mission. Support for the operation of Hinode is funded by the U.K. Space Agency, which works with the Science and Technology Facilities Council to fund researchers who exploit the scientific data yielded by the mission. NASA provided telescope components and major EIS optical components.

For more information on Hinode, please visit *http://www.nasa.gov/mission_pages/solar-b/*.

Science

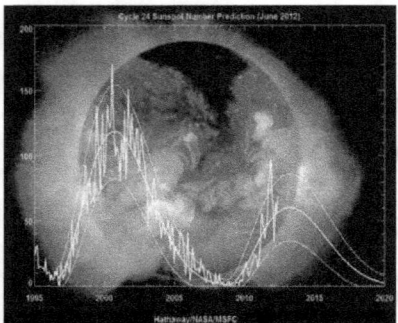

This image shows sunspot predictions for Solar Cycle 24. The space-radiation environment poses a hazard to spacecraft, which must therefore protect their sensitive components with adequate radiation shielding. Planning for satellite orbits and space missions often requires knowledge of solar activity levels years in advance. The Space Environment Testbeds will improve hardware performance in the space-radiation environment.

United Kingdom

SET-1
Space Environment Testbed-1

The Space Environment Testbed–1 (SET-1) project performs flight and ground investigations to address the Living With a Star (LWS) program's goal of understanding how interactions between the Sun and Earth affect humanity. The SET-1 project is the element of the LWS Program that characterizes the space environment and its impact on hardware performance in space. The project goal for SET-1 is to improve the engineering approach to accommodation and/or mitigation of the effects of solar variability on spacecraft design and operations. The United Kingdom (U.K.) Space Agency is collaborating with NASA on this mission. International participation also includes the French National Center for Space Studies (CNES).

The primary science objectives for the SET-1 project are to define space environment effects and mechanisms, to reduce design margins, and to improve design and operations guidelines. When margins for spacecraft design and operations are reduced, "space-environment-tolerant" and new technologies can be used more frequently, the fraction of spacecraft resources available for payloads can be increased or launch vehicle requirements can be reduced, and routine operations above low-Earth orbit (LEO) can be performed at the same cost as for LEO operations.

NASA and the U.K. Space Agency are conducting a British investigation entitled "Cosmic Radiation Environment Dosimetry and Charging Experiment" (CREDANCE) as part of the SET-1 project. The investigation uses a correlative environment monitor (CEM) to measure the energetic space-radiation environment during the mission. The CEM measurements will support the data analysis for other experiments on the SET-1 project and can be used to improve and/or validate environment specification models for electrons, protons, and heavy ions, and to predict solar particle events.

The U.K. Space Agency will make available an experiment carrier that can be used by the CREDANCE CEM, and it will provide the interface between the CEM and the host spacecraft. SET-1 will be flown on the Air Force Research Laboratory (AFRL) Demonstration and Science Experiments (DSX) spacecraft and is scheduled to launch in 2014.

For more information on SET, please visit *http://science.nasa.gov/missions/space-environment-testbeds/*.

United Kingdom

Swift Gamma-ray Burst Explorer

NASA's Swift Gamma Ray Burst Explorer mission is a multiwavelength astrophysics observatory that is making a comprehensive study of hundreds of gamma-ray bursts (GRBs) in order to determine the origin and physical processes of GRB events. The Swift mission was launched by NASA in November 2004 and was expected to have an orbital lifetime of approximately 8 years. Updated orbital lifetime predictions for Swift project it to remain in orbit potentially until 2022. The United Kingdom (U.K.) Space Agency and the Italian Space Agency (ASI) are cooperating with NASA on the operation of the Swift mission.

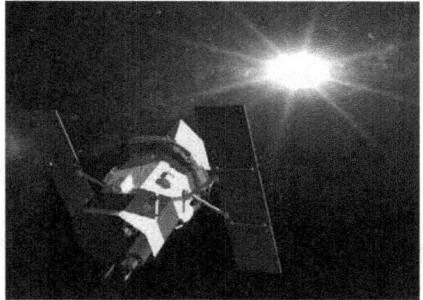

An artist's rendering of NASA's Swift spacecraft.

GRBs are incredibly intense releases of gamma radiation, which is a particularly energetic form of light that can only be generated by the most powerful astronomical events. Swift is a next-generation satellite that observes GRBs—a first-of-its-kind, multiwavelength observatory dedicated to advancing GRB science. Its three instruments work together to patrol about one-sixth of the sky at a time in order to pinpoint and observe GRBs and afterglows in the gamma-ray, x-ray, ultraviolet, and optical wavebands.

Scientific institutes in the United Kingdom cooperated with NASA on both the development of the X-ray Telescope (XRT) and on the assembly of the Ultraviolet and Optical Telescope (UVOT), having supplied system engineering support and expertise for the XRT design; provided the XRT structural finite element model, thermal finite element model, focal plane camera assembly, and other major components for the XRT; and developed the XRT and UVOT data analysis software.

For more information on Swift and the XRT, please visit *http://swift.gsfc.nasa.gov/* and *http://www.nasa.gov/mission_pages/swift/spacecraft/index.html*.

This composite image, taken from three Swift UVOT ultraviolet images, highlights the luminous hot gas in the Crab Nebula, a supernova remnant. (Credit: NASA, Penn State University/Erik Hoversten)

WORLDWIDE MISSIONS AND PROGRAMS

Sun photometers measure the intensity of sunlight arriving directly from the Sun. They are pointed directly at the Sun and measure direct sunlight. Since haze and aerosols block some direct sunlight, a Sun photometer is an ideal instrument for measuring haze.

Close up of AERONET's Sun photometer on Bhola Island in Bangladesh.

Worldwide

AERONET
Aerosol Robotic Network

Aerosol particles come from both human and natural sources, can be observed everywhere in the world, and play a key role in climate. The Aerosol Robotic Network (AERONET) is an optical, ground-based, aerosol-monitoring network of Sun photometers and data archive system supported by NASA and expanded by a federation of over 80 countries and regions. Data from this collaboration provide globally distributed, near-real-time observations of spectral aerosol optical depths, aerosol size distributions, and precipitable water in diverse aerosol environments.

By providing accurate measurements from the ground, AERONET has emerged as the best tool to validate the accuracy of satellite instruments. For example, scientists have relied upon AERONET to reconcile differences between aerosol measurements from the Moderate Resolution Imaging Spectroradiometer (MODIS) and the Multiangle Imaging SpectroRadiometer (MISR), two instruments on NASA's Terra satellite.

The data undergo preliminary processing (real-time data), reprocessing (final calibration approximately 6 months after data collection), quality assurance, archiving, and distribution from NASA's Goddard Space Flight Center master archive database and from several identical databases maintained globally. The data provide algorithm validation of satellite aerosol retrievals, as well as characterization of aerosol properties unavailable from satellite sensors.

NASA has agreements to support AERONET Sun photometer sites in 36 countries: Bangladesh, Bermuda, Bolivia, Colombia, Estonia, France, India, Indonesia, Israel, Japan, Kazakhstan, Kenya, Malaysia, Moldova, Mongolia, Morocco, Mozambique, New Zealand, Norway, Pakistan, Peru, the Philippines, Poland, Russia, Saudi Arabia, Singapore, South Africa, South Korea, Spain, Taiwan, Thailand, Turkey, Uganda, the United Arab Emirates, the United States, and Vietnam.

AERONET Sun photometer sites are independently operated in an additional 55 locations. The AERONET sites in these locations, while not directly supported by NASA, are tied into the network and their data is openly shared: Amsterdam Island, Antarctica, Argentina, Ascension Island, Australia, Austria, the Azores, Barbados, Belarus, Belgium, Brazil, Burma, Canada, Cape Verde, China, Crozet Island, Cuba, Cyprus, Denmark, Egypt, Finland, French Polynesia, Germany, Guadeloupe, Guam, Guyane, Iran, Italy, Kuwait, Kyrgyzstan, Laos, Maldives, Mali, Mauritania, Mauritius, Mexico, Midway Islands, North Mariana Islands, Nauru, Nepal, Netherlands, Niger, Nigeria, Papua New Guinea, Portugal, Puerto Rico, Romania, Senegal, Sweden, Switzerland, Tajikistan, Tenerife Island, Ukraine, the United Kingdom, and Zambia.

For more information on AERONET, please visit *http://aeronet.gsfc.nasa.gov/*.

Worldwide

AMS
Alpha Magnetic Spectrometer

The AMS project addresses fundamental issues shared by physics, astrophysics, and cosmology on the origin and structure of the universe. The visible matter in the universe (i.e., stars) composes up to less than 5 percent of the total mass that is known to exist. The other 95 percent is dark—either dark matter, which is estimated at 20 percent of the universe by weight, or dark energy, which makes up the balance. AMS specifically looks for dark matter, as well as antimatter, which in the Big Bang theory of the origins of the universe is required to exist in equal parts to matter.

A computer-generated drawing of the AMS-02.

The first AMS results, released in April 2013, provided the most precise measurement to date of the ratio of positrons to electrons in cosmic rays. Measurements of this key ratio may eventually provide our first glimpse into the nature of dark matter. NASA Administrator Charles Bolden noted that the results "could help foster a new understanding of the fields of fundamental physics and astrophysics."

The AMS project is led by the U.S. Department of Energy. The principal investigator is Nobel-laureate physicist Samuel Ting. The AMS project is composed of an international team from the following 16 countries: China, Denmark, Finland, France, Germany, Italy, Mexico, the Netherlands, Portugal, Romania, Russia, South Korea, Spain, Switzerland, Taiwan, and the United States. The European Space Agency (ESA) and the European Organization for Nuclear Research (CERN) are also members of the AMS team.

The AMS-02 after installation on the International Space Station's (ISS) S3 Upper Inboard Payload Attach Site on the S3 Truss segment. The AMS-02 will remain active for the duration of the ISS Program.

For more information on AMS, please visit *http://ams.nasa.gov/*.

Human Exploration and Operations

Worldwide

Astrobiology and Analogs

Astrobiology is a multidisciplinary field of research that encompasses studies of the origin, evolution, distribution, and future of life in the universe. Research in exobiology and evolutionary biology ranges from the molecular biology of the origin of life, to the biochemistry of adaptation to extreme environments, to studies of fossil life on Earth, to prebiotic chemistry on worlds such as Titan. The NASA Astrobiology Program is developing and demonstrating sensors, robotics, and human exploration techniques and technologies in terrestrial analog environments for use in space-based astrobiological studies and related investigations.

NASA astrobiologists collaborate with astrobiologists around the world on data analysis, field research, flight experiments, mission planning, and more. Field sites for astrobiology research have ranged from Antarctica to Alaska, Australia, Canada, Chile, China, Hawaii, Mexico, and Norway, as well as elsewhere in the continental United States. Many of these sites also provide value beyond the realm of astrobiology by serving as "analogs" to destinations of interest to future human explorers—the Moon, Mars, and asteroids, for example. Analog missions are conducted at these often remote locations due to their physical similarities to the extreme environments associated with these exploration destinations. NASA engineers and scientists work with international partners at these sites to test robotic equipment, vehicles, habitats, communications, and power systems.

The NASA Astrobiology Institute (NAI) has supported the work of U.S. investigators whose investigations are part of the plan for the European Space Agency's (ESA) ExoMars mission to study the biochemical environment on Mars. Formally, the NAI has 13 international partners: the European Astrobiology Network Association, the Nordic Network of Astrobiology, and national astrobiology research institutes in Australia, Brazil, Canada, Colombia, France, Germany, the United Kingdom, Mexico, Russia, and Spain. U.S. scientists on NAI research teams are collaborating with researchers at universities in Athens, Leeds, Leiden, Oslo, Paris, Taipei, Tokyo, and Toronto, as well as scientists at the Vatican Observatory and the Brazilian Space Agency (AEB). NASA also collaborates with astrobiologists from other nations through international organizations such as the Committee on Space Research (COSPAR) of the International Council for Science.

In attempting to learn about the origin and evolution of life, astrobiologists studying microbial communities in extreme environments on Earth have discovered that the limits within which terrestrial life can exist are far broader and that the diversity of microbial life is far greater than scientists previously thought. These findings are contributing to the important field of genomics (the study of genes and their functions), aiding planning for experiments and missions to search for evidence of possible life on Mars and other planetary bodies, and generally improving our understanding of the nature and diversity of life on Earth and potentially elsewhere.

For more information on the Astrobiology program, please visit *http://astrobiology.nasa.gov/*.

Titan's heavy atmosphere is loaded with organic molecules that may provide insights on the evolution of Earth's early environment and the emergence of life.

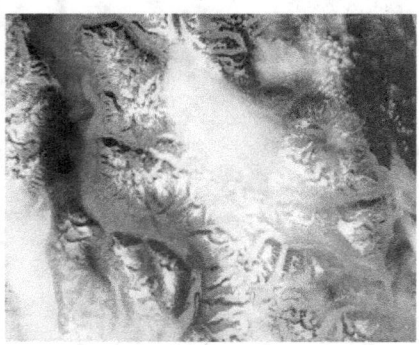

Far above the Arctic Circle off the northern coast of Norway lies a small chain of islands known as Svalbard. The rugged beauty, ancient fossil beds, and volcanic past of these islands make them an ideal place to learn how to explore other worlds for signs of life. The Advanced Spaceborne Thermal Emission and Reflection Radiometer (ASTER) on NASA's Terra satellite captured this false-color image of the region in June 2001.

Worldwide

Balloons

Large unmanned helium balloons provide NASA with an inexpensive means to place scientific and technology payloads into a near-space environment at a fraction of the cost of a satellite mission. In addition, balloon missions can be prepared for flight in as little as 6 months. The unique capabilities of this program are crucial for the development of new technologies and payloads for NASA's space flight missions. Many important scientific observations in fields such as hard x-ray, gamma-ray, and infrared astronomy; cosmic rays; atmospheric studies; and others have been made from balloons.

Balloons play a major role in the education and training of young scientists and engineers. It is possible for undergraduate and graduate students to design and conduct a balloon science mission within the average time of a graduate degree program. Many of the principal investigators and engineers leading NASA's space missions got their start in the balloon program.

NASA currently flies conventional and long-duration balloon missions. Typically, a conventional balloon flight will last from 6 to 36 hours, while a long-duration flight can last up to several weeks. In 2009, NASA successfully launched and demonstrated a newly designed super-pressure balloon that is capable of flying at constant altitude for mission durations up to 100 days at any geographic latitude.

Balloon launches occur from permanent launch sites in Palestine, Texas, and Fort Sumner, New Mexico. NASA also uses sites in Sweden, Australia, and Antarctica. Presently, NASA has a treaty with the government of Australia for a balloon launch site in Alice Springs, Australia, and has concluded agreements for launches from Sweden's ballooning facility in Esrange, Sweden. The scientific balloon payloads that NASA launches are built by and gather scientific data for use by teams that include members from universities around the globe.

The NASA Balloon Program Office is located at the Goddard Space Flight Center (GSFC) Wallops Flight Facility (WFF).

For more information on NASA's Balloon Program, please visit *http://www.nasa.gov/centers/wallops/home/organization.html*.

A NASA high-altitude, long-duration balloon carrying the Balloon-borne Large Aperture Sub-millimeter Telescope (BLAST) prepares for launch at McMurdo Station in Antarctica. By providing sensitive, large-area, sub-millimeter surveys, BLAST helps address important questions regarding the formation and evolution of stars, galaxies, and clusters. (Credit: BLAST)

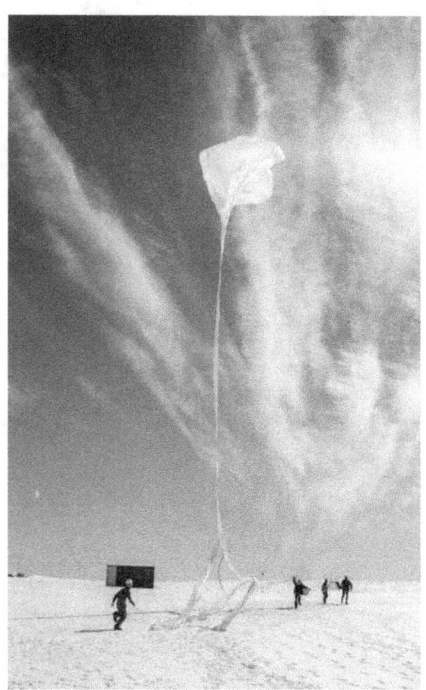

Scientists race through the Antarctic snow to launch 1 of 20 balloons as part of NASA's Balloon Array for Radiation belt Relativistic Electron Losses (BARREL) mission. Each balloon is equipped with instruments to help track how electrons from giant radiation belts surrounding Earth travel down magnetic field lines toward the poles. BARREL works in conjunction with NASA's Van Allen Probes.

Education and Outreach

Teams compete in the annual Great Moonbuggy Race.

With students and Leland Melvin (NASA astronaut and NASA Associate Administrator for Education) looking on, musical artist will.i.am posts a tweet shortly after his song "Reach for the Stars" was beamed back from the Curiosity Mars rover and broadcast to a live audience at NASA's Jet Propulsion Laboratory in Pasadena, California. (Credit: NASA/Jet Propulsion Laboratory–Caltech)

Worldwide

Education

Students and educators from around the world partake in NASA education programs and take advantage of NASA education resources to enrich their understanding of science, technology, engineering, and mathematics (STEM) subject matter. NASA utilizes unique assets like the International Space Station (ISS) to inspire the next generation of scientists and explorers with such programs as the Amateur Radio on the International Space Station, ISSLive!, and Earth Knowledge Acquired by Middle School Students (EarthKAM). Students exercise in the footsteps of space's elite explorers through NASA's Mission X: Train Like an Astronaut program, which encourages students to develop good health and fitness habits, while at the same time teaching students important STEM concepts.

NASA maximizes technology to reach as broad an audience as possible. One such example is NASA's Digital Learning Network (DLN), which links classrooms from around the world in virtual classes with subject matter experts who teach a variety of STEM topics.

Carefully crafted design competitions challenge students to solve engineering problems while working together to excel as a team. Annual events like the Lunabotics Mining Competition and the Great Moonbuggy Race bring students from around the world to challenge and showcase their ingenuity.

NASA recognizes the importance of international collaboration in education and actively participates in a number of international forums. NASA is a founding member of the International Space Education Board, an active member of the International Astronautical Federation's Space Education and Outreach Committee, a strong supporter of the International Space University, and the lead on the ISS Program Science Forum's Working Group on Education. Through internships, NASA is able to provide a platform where U.S. students and students from around the world can work together to learn how collaboration can lead to innovative solutions.

NASA's educational materials are readily available on the NASA Web site or via the Central Operation of Resources for Educators.

A sample of NASA educational activities with international participation include: Amateur Radio on the International Space Station; Green Aviation University Engineering Contest; Digital Learning Network; EarthKAM; Gravity Recovery and Interior Laboratory (GRAIL) Moon Knowledge Acquired by Middle School Students (MoonKAM); Great Moonbuggy Race; NASA Aeronautics High School Contest; International Observe the Moon Night; ISSLive!; Lunabotics Mining Competition; Mission X: Train Like an Astronaut; Museum Alliance; Spaceward Bound; and Zooniverse.

For more information on NASA's education program, please visit *http://www.nasa.gov/education*.

Worldwide

Exhibits

NASA, on a limited basis due to cost and other constraints, provides and supports NASA- and space-themed exhibits around the world. These exhibits, which typically attract great interest, help educate and inform audiences of NASA's programs, missions, and discoveries. Because it receives many more requests for exhibits than it can accommodate, NASA carefully evaluates each exhibit request and considers how NASA's participation in the particular event could contribute to the Agency's mission. NASA often coordinates with the U.S. Department of State to evaluate requests and identify broader U.S. Government interests that could be advanced through NASA's provision of an exhibit. In some cases, NASA proactively offers to provide an exhibit and/or conduct outreach activities if doing so furthers the interests of NASA and the United States.

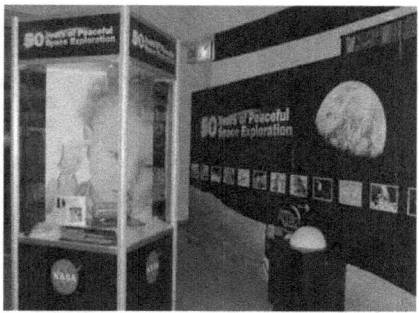

NASA's exhibit at the United Nations in Vienna, Austria, during its celebration of 50 years of peaceful space exploration.

There are a number of ways NASA can support an event, depending upon the nature of the event and the venue. These options include: the use of publicly available NASA media, including photographs, images, pop-up displays, science data, and audio and video recordings; the loan of a moon rock, special artifact, or model; and/or on-site appearances and assistance by NASA personnel. Standard procedures require that the production, shipping, insurance, security measures, and other exhibit-related overhead and logistics meet NASA standards and are funded by the exhibit sponsors.

A look inside the NASA: A Human Adventure exhibition by John Nurminen Events.

Some private companies can provide space-themed exhibits. A family-owned company in Finland, called John Nurminen Events, has developed what is considered to be the world's largest touring space exhibition, entitled NASA: A Human Adventure. This exhibition, which is geared toward international audiences, is focused on the history and value of space exploration. It covers over 2,000 square meters and displays over 250 original objects and replicas. Included in the collection are space capsules, rocket engines, astronaut suits, space telescopes, and satellites. NASA entered into an international agreement with John Nurminen Events in 2010 to facilitate NASA participation in the exhibition. The exhibition premiered in Stockholm, Sweden, in January 2011; traveled next to Madrid, Spain, in December 2011; and then opened in Istanbul, Turkey, in September 2012. The exhibition will travel to several more cities in the coming years.

For more information on NASA exhibits, please visit *http://www.nasa.gov/about/exhibits/index.html*, *http://www.ahumanadventure.com/*, and *http://www.nasa.gov/about/exhibits/indexFeature-RequestExhibit.html*.

Science

Worldwide

FEWS-NET
Famine Early Warning System Network

A 2008 MODIS image of winter snowpack (right) shows that the snowpack was significantly lighter than in 2007 (left) because of the prevailing dry weather pattern across Afghanistan. (Credit: NASA/MODIS, U.S. Department of Agriculture)

The Famine Early Warning System Network (FEWS-NET), operated by the U.S. Geological Survey (USGS), collaborates with international, regional, and national partners to provide timely and rigorous early warning and vulnerability information on emerging and evolving food security issues. FEWS-NET professionals in Africa, Central America, Haiti, Afghanistan, and the United States monitor and analyze relevant data and information in terms of their impact on livelihoods and markets to identify potential threats to food security.

Once these issues are identified, FEWS-NET uses a suite of communications and decision support products to help decision makers act to mitigate food insecurity. These products include monthly food security updates for 25 countries, regular food security outlooks and alerts, and briefings and support for contingency- and response-planning efforts. FEWS-NET also focuses its efforts on developing capacity, building and strengthening networks, developing useful policy information, and building consensus around food-security problems and solutions.

NASA satellites and instruments (such as Terra, Aqua, the Tropical Rainfall Measuring Mission, and the Moderate Resolution Imaging Spectroradiometer [MODIS]), as well as other products (such as the NASA Land Atmosphere Near-real-time Capability for Earth Observing System [EOS] [LANCE]), support these efforts by providing continuously updated imagery of affected areas. MODIS Land Surface Temperature data are used to calculate actual evapotranspiration values over the entire African continent.

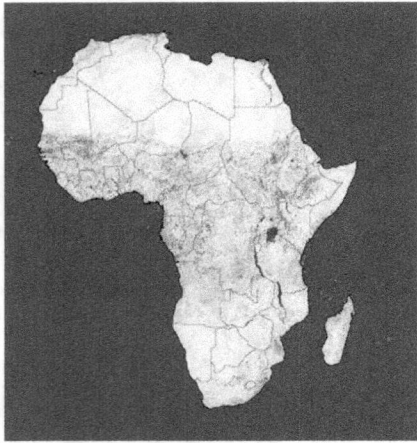

This September 2002 Normalized Difference Vegetation Index (NDVI) image depicts vegetation density across Africa. Areas where vegetation density is less than the 20-year average are colored brown, while above-average density is colored green. Such images are used to track drought conditions that may threaten subsistence agriculture, which is the primary food source for populations around the world. (Credit: NASA/MODIS)

Products from this system were used to help support the formal declaration of famine in Somalia by the United Nations and the U.S. Agency for International Development (USAID) in July 2011.

For more information on FEWS-NET, please visit *http://www.fews.net/Pages/default.aspx*.

Worldwide

GES
Global Exploration Strategy

In early 2006, NASA initiated a dialogue with representatives of 13 science and space agencies from around the world with the goal of coordinating a global strategy for exploration. Following a year of focused discussions, NASA—along with organizations representing Australia, Canada, China, the European Space Agency (ESA), France, Germany, Great Britain, India, Italy, Japan, Russia, the Republic of Korea, and Ukraine—released "The Global Exploration Strategy: The Framework for Coordination." This framework document, which is the product of a shared vision of space exploration focused on solar system destinations where humans may someday live and work, represented an important first step in coordinating space exploration efforts toward common goals.

The authors of the framework document recognized the need for a voluntary, nonbinding mechanism by which space agencies could exchange information on their respective space exploration plans. In 2007, the participating agencies established terms of reference for an International Space Exploration Coordination Group (ISECG). The ISECG plays a key role in helping to identify gaps, overlaps, and synergies in the space exploration plans of participating agencies.

In 2007, 14 science and space agencies from around the world released "The Global Exploration Strategy: The Framework for Coordination."

Building on the Global Exploration Strategy's vision to coordinate human and robotic exploration of our solar system, in 2011, the participating agencies of the ISECG developed and released the first iteration of the "ISECG Global Exploration Roadmap." This document reflects the international effort to define feasible and sustainable exploration pathways to near-Earth asteroids, the Moon, and Mars, and creates a framework for greater international exploration discussions. The roadmap demonstrates the importance of the International Space Station as a first step and a bridge to the exploration of destinations beyond low-Earth orbit. The roadmap will be updated over time to reflect evolving global consensus on exploration destinations and priorities. The latest updated version illustrates planned and conceptual near-term missions that advance human and robotic exploration starting in the Earth-Moon system. It also takes into account innovative ideas and concepts from external stakeholders since the roadmap was first issued in September 2011.

The release of the framework document, the subsequent establishment of the exploration coordination group, and the release of the first Global Exploration Roadmap represent important steps in an evolving process toward a comprehensive global approach to space exploration. Although this multilateral effort is nonbinding, the process is consistent with ongoing bilateral and multilateral discussions that NASA hopes will lead to cooperative agreements for specific projects.

For more information on the ISECG, the Global Exploration Roadmap, or the Global Exploration Strategy, please visit *http://www.globalspaceexploration.org/home*.

Science

Worldwide

GLOBE
Global Learning and Observations to Benefit the Environment

The Global Learning and Observations to Benefit the Environment (GLOBE) program is an international program that was initiated on Earth Day in 1995 as a hands-on, primary- and secondary-school-based science and education program that unities students, teachers, scientists, and community members around the world in studying and conducting research about Earth's environment, connecting the local perspective to the vantage from space. Bilateral agreements establish partnerships between the United States and international partner nations. International partners sponsor GLOBE activities in their countries, designing and funding their own implementation strategies to be compatible with their national and regional educational priorities.

NASA leads the GLOBE program in close partnership with the U.S. National Science Foundation and the U.S. National Oceanic and Atmospheric Administration (NOAA), with support from the U.S. Department of State.

There are 109 participating countries in the GLOBE program. Since its inception in 1995, over 1.5 million students in more than 24,000 schools have participated in the program. GLOBE students take important environmental measurements focusing on atmosphere and climate, hydrology, soils, land cover biology, and phenology. GLOBE students have entered over 23 million environmental measurements into the international database on the GLOBE Web site. This data is freely available for use by students and scientists worldwide. By involving students in scientific data collection and research, including taking measurements, analyzing data, and participating in research collaborations with other students, GLOBE provides students with a full and practical understanding of the scientific process.

The following 109 countries participate in the GLOBE program: Argentina, Austria, Bahamas, Bahrain, Bangladesh, Belgium, Benin, Bolivia, Bulgaria, Burkina Faso, Cameroon, Canada, Cape Verde, Chad, Chile, Colombia, Congo, Costa Rica, Croatia, Cyprus, Czech Republic, Denmark, Dominican Republic, Ecuador, Egypt, El Salvador, Estonia, Ethiopia, Fiji, Finland, France, Gabon, Gambia, Ghana, Greece, Guatemala, Guinea, Honduras, Hungary, Iceland, India, Ireland, Israel, Italy, Japan, Jordan, Kazakhstan, Kenya, Kuwait, Kyrgyzstan, Latvia, Lebanon, Liechtenstein, Lithuania, Luxembourg, Macedonia, Madagascar, Maldives, Mali, Malta, Marshall Islands, Mauritania, Mexico, Micronesia, Moldova, Monaco, Mongolia, Montenegro, Morocco, Namibia, Nepal, Netherlands, New Zealand, Niger, Nigeria, Norway, Oman, Pakistan, Palau, Panama, Paraguay, Peru, Philippines, Poland, Portugal, Qatar, Romania, Russia, Rwanda, Saudi Arabia, Senegal, Serbia, South Africa, South Korea, Spain, Sri Lanka, Suriname, Switzerland, Tanzania, Thailand, Trinidad and Tobago, Tunisia, Turkey, Uganda, Ukraine, United Arab Emirates, the United Kingdom, the United States, and Uruguay.

For more information on GLOBE, please visit *http://www.globe.gov/*.

GLOBE students in Benin log data from a rain gauge.
(Credit: GLOBE)

GLOBE students in their classroom in Nepal, India.
(Credit: GLOBE)

Worldwide

International Astronauts

NASA works with numerous international space agencies, including the Canadian Space Agency (CSA), the European Space Agency (ESA), the Japan Aerospace Exploration Agency (JAXA), and the Russian Federal Space Agency (Roscosmos), to assemble international astronaut crews. During the 135 Space Shuttle missions, 63 international astronauts and cosmonauts from 15 different countries flew on the Space Shuttle 89 times. Since November 2000, International Space Station (ISS) Expedition crews have continuously inhabited the ISS. Astronauts, cosmonauts, and non-astronaut space flight participants from 15 different countries (including the United States) have visited the ISS.

The six-person Expedition 20 crew poses in "star-burst" formation for an in-flight portrait in the Harmony node of the International Space Station. Pictured clockwise from right (center) are: Russian cosmonaut Gennady Padalka, Canadian astronaut Robert Thirsk, Japanese astronaut Koichi Wakata, U.S. astronaut Michael Barratt, Russian cosmonaut Roman Romanenko, and ESA astronaut Frank De Winne.

The Multilateral Crew Operations Panel (MCOP)—consisting of representatives from ESA, JAXA, CSA, Roscosmos, and NASA—defines the processes, standards, and criteria for selection, assignment, training, and certification of an ISS crew for flight. The MCOP typically assigns ISS Expedition mission crews 2½ years prior to their mission.

Each ISS Expedition crew consists of dedicated astronauts and cosmonauts who train together to prepare for their missions. The ISS Expedition crew's training regimen takes crewmembers around the world as they travel to training facilities at each ISS partner agency to prepare for the wide variety of tasks that they will need to perform while on orbit. Crewmembers receive training in a wide variety of areas, including extra-vehicular activity (EVA), robotics, ISS systems, payloads, Soyuz spacecraft, leadership and human behavior, survival skills, and foreign languages. Crewmembers also complete more in-depth, specialized training for the particular role that he or she has been assigned to fill during the mission.

For more information on astronauts, please visit *http://www.nasa.gov/astronauts/* and *http://www.jsc.nasa.gov/Bios/index.html*.

Science

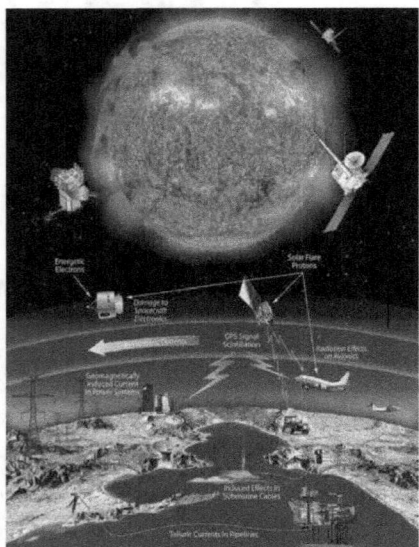

Technological infrastructure affected by space weather events includes satellites, aircraft, and power grids. A web of interdependencies makes the modern economy especially sensitive to solar storms. This is why advancing our understanding of the causes of space weather and improving its forecasting are critical goals.

Worldwide

ISWI
International Space Weather Initiative

The International Space Weather Initiative (ISWI) is a program of international cooperation to advance space weather science through the establishment of a distributed network of globally deployed science instruments that are used to collect space weather data for analysis and interpretation. ISWI was initiated as a follow up to the International Heliophysical Year of 2007, with a focus strictly on space weather. The goal of ISWI is to develop the scientific insight necessary to understand space weather science and to reconstruct and forecast near-Earth space weather. This includes instrumentation, data analysis, modeling, education, training, and public outreach.

ISWI seeks to establish and coordinate space weather monitoring stations across the globe to fill in gaps of observations of the near-Earth environment and of Earth's atmosphere, and also to encourage heliophysics research and education in developing nations. In conjunction with the United Nations, ISWI has installed and coordinated nearly 1,000 instruments located in 100 countries around the world.

The global network of these instruments provides scientists with constant viewing of both the Sun and the ionosphere, which has helped to spark the writing of numerous scientific papers and inspired the establishment of new graduate school programs across the world. This growth in space research and graduate programs has helped to increase the number of space scientists around the world and fostered international partnerships, which was the main motivation for the United Nations' involvement with ISWI.

Recent monitoring of the ionosphere by scientists in Algeria, Libya, and other countries in the ISWI network revealed that the ionosphere over that region is unusually active even in otherwise quiet times, leading to increased interruption of radio communications. This information could only have been determined by observing numerous communication signals traveling different paths through the same disturbance.

For more information on ISWI, please visit *http://iswi-secretariat.org/*.

Worldwide

International Space Station Utilization: Science and Technology Research and Development

The International Space Station (ISS) is the only multinational space-based research and technology test bed, and it provides an unparalleled capability for research that cannot be pursued on Earth. The three major science laboratories aboard the ISS—the U.S. Destiny, European Columbus, and Japanese Kibo facilities—as well as external test beds and observatory sites enable astronauts to conduct a wide variety of experiments in the unique microgravity and ultra-vacuum environment of low-Earth orbit.

NASA is carrying out a number of science and technology research and development activities on the ISS in collaboration with agencies from around the world. Two examples of such collaboration include:

Capillary Channel Flow Experiment—Germany

Capillary Channel Flow (CCF) is a versatile experiment for studying a variety of inertial-capillary-dominated flows that are key to spacecraft systems and that cannot be studied on the ground. The results of CCF help innovate existing applications, and inspire new ones, in the field of the aerospace community that faces the challenge of containment, storage, and handling of large liquid inventories (e.g., fuels, cryogens, and water) aboard spacecraft. The results will be immediately useful for the design, testing, and instrumentation needed to verify and validate the liquid management systems of current orbiting, design-stage, and advanced spacecraft that are envisioned for future lunar and Mars missions. The results will also be used to improve life-support-system design and phase separation and to enhance current system reliability.

Astronaut Sunita Williams, International Space Station Expedition 15's flight engineer, performs one of the multiple tests of the Capillary Flow Experiment (CFE) investigation in the Destiny Laboratory of the ISS.

For more information on CCF, please visit *http://issresearchproject.grc.nasa.gov/MSG/CCF/*.

Bisphosphonates Study—Japan

One of the most significant medical concerns for long-duration human space flight is the potential for bone mass loss. Astronauts can experience regional losses of 1 to 2 percent per month while in space. The Bisphosphonates Study tests the effectiveness of bisphosphonates—a group of antiresorptive agents that block the breakdown of bone and are used to treat osteoporosis and other disorders related to bone turnover—as a countermeasure to space-flight-induced bone loss. If shown to be effective, bisphosphonates or other antiresorptive agents could help prevent several bone-related problems for crewmembers on the ISS and on future long-duration missions. These problems include loss of bone mineral mass and the possibility of developing renal stones during or after space flight.

For more information on the Bisphosphonates Study, please visit *http://www.nasa.gov/mission_pages/station/research/experiments/Bisphosphonates.html*.

Science

Dan Irwin, SERVIR Program Director, NASA Marshall Space Flight Center (fifth from right), and Nancy Searby, NASA Applied Sciences Capacity Building Program Manager (fourth from right), stand with SERVIR team members from the Nairobi, Kenya, hub as they display Malawi and Rwanda land cover maps developed in a SERVIR Greenhouse Gas Inventories project. Pictured from left to right are: Hilda Manzi, Stephen Karanu, Fred Ogor Mokua, Andre Kooiman, Dan Irwin, Nancy Searby, Viola Kirui, Ababu Jaffer, and Phoebe Oduor. (Credit: SERVIR)

Canadian astronaut Chris Hadfield with the ISERV camera in Destiny Module's Earth-facing window, shortly after ISERV was delivered to the ISS.

Worldwide

SERVIR

SERVIR is a joint program between NASA and the U.S. Agency for International Development (USAID) that provides satellite-based Earth observation data and science applications to help developing nations in Central America, East Africa, and the Himalayas improve their environmental decision making.

SERVIR, which means "to serve" in Spanish, uses images and data from Earth-observing satellites and a new camera system on the International Space Station (ISS) called the ISS SERVIR Environmental Research and Visualization (ISERV) system to provide critical information about wildfires, floods, landslides, air quality, extreme weather, biodiversity, and changes in land cover. This information helps countries to assess environmental threats and respond to damage from natural disasters. In a very real sense, SERVIR provides basic information for living on Earth.

SERVIR began in 2004 as a collaborative effort among NASA, USAID, the World Bank, and the Central American Commission for Environment and Development (CCAD). The program, managed at NASA's Marshall Space Flight Center (MSFC) in Huntsville, Alabama, opened its first "hub" in Panama in 2005 to serve Central America and the Dominican Republic in cooperation with the Water Center for the Humid Tropics of Latin America and the Caribbean (CATHALAC). Additional hubs were opened in 2008 in Nairobi, Kenya, serving all of East Africa, and in 2010 in Kathmandu, Nepal, serving the Kush-Himalaya region. These hubs are partnerships with the Regional Center for Mapping of Resources for Development in Nairobi and the International Centre for Integrated Mountain Development in Kathmandu, respectively. Additional hubs are under consideration for other developing regions of the world.

The SERVIR hubs consist of teams of scientific experts who translate satellite data into useful information for governments and non-governmental organizations in the host nations. SERVIR's objective is to make each hub an integral, self-sustaining part of the national and regional information infrastructure. The SERVIR Coordination Office at the Marshall Space Flight Center provides scientific expertise and technologies to help the hubs advance their knowledge and capabilities. USAID's expertise in international development, training, and capacity-building helps bring NASA expertise and assets to bear on real-world problems by, as they say at SERVIR, "connecting space to village."

For more information on SERVIR, please visit *https://servirglobal.net/Global/About.aspx*.

Worldwide

Sounding Rockets

The NASA Sounding Rocket Program (NSRP) is a suborbital space flight program that supports NASA-sponsored science research activities, other U.S. agencies, and international sounding rocket groups and scientists. The approximately 20 suborbital missions flown annually by NSRP provide researchers with unparalleled opportunities to build, test, and fly new instrument and sensor design concepts while simultaneously conducting world-class scientific research. The short mission life cycle of sounding rockets, coupled with a hands-on approach to instrument design, integration, and flight, helps ensure that the next generation of space scientists receive the training and experience necessary to move on to NASA's larger, more complex space science missions. The cost structure and risk posture under which the program is managed stimulates innovation and technology maturation and enables rapid response to scientific events.

With the capability to fly higher than many low-Earth orbiting satellites and the ability to launch on demand, sounding rockets offer, in many instances, the only means to study specific scientific phenomena of interest to many researchers. Unlike the instruments on board most orbital spacecraft or in ground-based observatories, sounding rockets can place instruments directly into regions where and when the phenomena is occurring to enable direct, in situ measurements. The mobile nature of the program enables researchers to conduct missions from strategic vantage points worldwide. Telescopes and spectrometers to study solar and astrophysics are flown on sounding rockets to collect unique science data and to test prototype instruments for future satellite missions. An important aspect of most satellite missions is the calibration of space-based sensors. Sounding rockets offer calibration and validation flights for many space missions, particularly solar observatories such as NASA's Solar Dynamics Observatory (SDO).

NASA's Wallops Flight Facility (WFF), located on Virginia's eastern shore, is the only facility in the United States that designs, manufactures, fabricates, integrates, tests, and launches sounding rockets for scientific research. In addition to launching sounding rockets from WFF, NASA's Sounding Rocket Program launches these rockets from a variety of sites throughout the world, including Australia, Brazil, Canada, Greenland, Kenya, Norway, the Marshall Islands, and Peru.

For more information on NASA's Sounding Rocket Program, please visit *http://sites.wff.nasa.gov/code810/*.

NASA successfully tested a Talos Terrier Oriole Sounding Rocket in September 2012 from the Agency's launch range at the Wallops Flight Facility in Virginia. It was the first flight of the 65-foot-tall Talos Terrier Oriole, which is being developed to support high-altitude space science research.

One of two Charge and Mass of Meteoric Smoke Particles (CHAMPS) sounding rockets that were launched in October 2011 from the Andoya Rocket Range in northern Norway. CHAMPS instruments provided the first measurements of both the number density and size of meteoric dust particles, which are the seeds for the bright "noctilucent" clouds that appear over the poles at night.

Worldwide

Space Geodesy

Space geodesy is the science of accurately measuring and understanding Earth's geometric shape, its orientation in space, and its gravity field. Geodetic observing systems and the global geodetic infrastructure underlie navigation systems and support a wide array of military, research, civil, and commercial applications.

The global space geodetic infrastructure consists of a network that combines four techniques—Satellite Laser Ranging (SLR), Very Long Baseline Interferometry (VLBI), the Global Positioning System (GPS), and the Doppler Orbitography and Radiopositioning Integrated by Satellite (DORIS) system—to define the terrestrial and celestial reference frames and Earth's orientation in space. Among the many important applications are the study of ocean circulation, sea level change, changes in Earth's ice caps, deformation leading to earthquakes, volcanic eruptions, and landslides. The space geodetic network also enables precision deep-space, air, and land navigation used in most scientific and commercial applications.

NASA cooperates with over 30 countries in space geodesy research in support of the Global Geodetic Observing System (GGOS). NASA also supports the International GPS Service (IGS), the International VLBI Service (IVS), and the International Laser Ranging Service (ILRS). These three organizations place geodetic network operations, analysis, and standards within an international framework of cooperation and collaboration under the aegis of the International Association of Geodesy.

As part of the cooperation, NASA often loans equipment to host countries for use in space geodetic ground stations. The host countries, in turn, provide utilities, security, housing, operation, and maintenance of the equipment. These ground stations provide critical geodetic reference points within the host country. The inclusion of the stations within the global geodetic infrastructure significantly improves the accuracy of measurements at all scales.

All preprocessed and analyzed NASA space geodetic data are available from the Crustal Dynamics Data Information System (CDDIS) at Goddard Space Flight Center (GSFC). The CDDIS also is a major data archive for the GGOS.

For more information on the space geodetic network, please visit *http://space-geodesy.nasa.gov/* and *http://science.gsfc.nasa.gov/solarsystem/sgp/*.

NASA's Next-Generation Satellite Laser Ranging (NGSLR) system, located at the Goddard Geophysical and Astronomical Observatory in Greenbelt, Maryland, is shown ranging to the Lunar Reconnaissance Orbiter orbiting the Moon. This process provides the information needed to model the onboard clock drift and allows high-precision orbit determination.

This 12-meter antenna is part of the next-generation geodetic Very Long Baseline Interferometry system at NASA's Goddard Geophysical and Astronomical Observatory in Greenbelt, Maryland. Essential new features include the fast antenna, broadband receiver, and high-rate recording at four settable bands.

MULTILATERAL REPRESENTATION AND COORDINATION

NASA is an active member of the international space, science, aeronautics, education, technology, and exploration communities. Through its participation in numerous international and multilateral committees, boards, and other groups, NASA's global reach extends beyond its specific missions and cooperative activities. NASA's active engagement in these groups helps it advance many important programmatic and policy priorities, such as sharing and coordinating the Agency's future plans, strategies, and science research priorities and initiatives; exchanging information and ideas on education and outreach initiatives; staying current on the latest cutting-edge science and technology developments in aeronautics and space; coordinating key Earth-science observations and activities; collaborating on the next steps in space exploration; and sharing and coordinating many other important areas of research, operations, education, and technology. Participation in these groups provides NASA with an opportunity to meet and discuss program and policy priorities not only with its longstanding and traditional partners but also with newer and potential nontraditional international partners.

The following pages contain a small sample of the international and multilateral groups in which NASA participates.

Multilateral Representation and Coordination

Committee on Earth Observation Satellites (CEOS): Established in 1984, the Committee on Earth Observation Satellites is the focal point for international coordination of space-borne Earth observation activities. CEOS currently focuses on key areas of climate, deforestation and forest carbon, agriculture and food security, disasters, water, and oceans to optimize societal benefits through the cooperation of members in mission planning and in the development of compatible data products, formats, services, applications, and policies. CEOS is the space component of the Global Earth Observation System of Systems (GEOSS), which facilitates coordinated, comprehensive, and sustained Earth observations and information that will inform decisions and actions for sustainable development, societal benefit, and an improved understanding of the Earth system. CEOS works closely with U.N. agencies in support of nine societal benefit areas and with other international organizations that complement its mission. For more information on CEOS, visit *http://www.ceos.org/*.

Committee on Space Research (COSPAR): Established in 1958 by the International Council of Scientific Unions, the Committee on Space Research provides the scientific community with a platform to promote scientific research in space on an international level, with particular emphasis on the exchange of results, information, and opinions. COSPAR also provides a forum, open to all scientists, for the discussion of problems that may affect scientific space research. These objectives are achieved through the organization of scientific assemblies, publications, and other means. For more information on COSPAR, visit *http://cosparhq.cnes.fr/about*.

Consultative Committee for Space Data Systems (CCSDS): Founded in 1982 by the major space agencies of the world, the Consultative Committee for Space Data Systems is a multinational forum for the development of communications and data-system standards for space flight. The CCSDS develops recommendations for data- and information-systems standards to promote interoperability and cross-support among cooperating space agencies and to enable multiagency space flight collaboration (both planned and contingency) and new capabilities for future missions. CCSDS standardization reduces the cost burden of space flight missions by allowing cost-sharing between agencies and cost-effective commercialization. For more information on CCSDS, visit *http://public.ccsds.org/default.aspx*.

Global Geodetic Observing System (GGOS): Established in 2003 as the observing system of the International Association of Geodesy (IAG), the Global Geodetic Observing System provides consistent observations of the spatial and temporal changes of the shape and gravitational field of Earth, as well as the temporal variations of Earth's rotation. GGOS represents IAG in the Group on Earth Observation (GEO) and is IAG's contribution to the Global Earth Observation System of Systems (GEOSS). For more information on GGOS, visit *http://www.iag-ggos.org/*.

Group on Earth Observations (GEO): Established following the 2002 U.N. World Summit on Sustainable Development, the Group on Earth Observations is a voluntary partnership of governments and international organizations that aims to build a Global Earth Observation System of Systems (GEOSS). The goal is to facilitate the coordination of future decisions and actions using comprehensive and sustained Earth observations and information in support of nine societal benefit areas: agriculture, biodiversity, climate, disasters, ecosystems, energy, health, water, and weather. GEO aims to improve interoperability, information-sharing, and the delivery of information to users. For more information on GEO, visit *http://www.earthobservations.org/index.shtml*.

Interagency Operations Advisory Group (IOAG): Created in 1999, the IOAG provides a forum to identify common needs across multiple international agencies for coordinating space communications policy, high-level procedures, technical interfaces, and other matters related to interoperability and space communications. The IOAG was established by the Inter-Operability Plenary, which was convened by six space agencies to encourage interagency interoperability for cross-support, in an effort to realize the additional economies resulting from an ability to share the large capital investments made by each agency in mission-support systems. For more information on IOAG, visit *https://www.ioag.org/*.

Interagency Space Debris Coordination Committee (IADC): Founded in 1993, the Interagency Space Debris Coordination Committee is an international forum for coordinating activities related to the issues of artificial and natural debris in space through the exchange of information on space debris research activities, the facilitation of opportunities for cooperation in space debris research, and the identification of debris mitigation options. For more information on IADC, visit *http://www.iadc-online.org/index.cgi*.

International Astronautical Federation (IAF): Created in 1951 to foster the dialogue between scientists around the world and to support international cooperation in all space-related activities, the International Astronautical Federation is a worldwide federation of organizations active in space that is responsible for the premier global space conference, the International Astronautical Congress (IAC). The IAF actively encourages the development of astronautics for peaceful purposes and supports the dissemination of scientific and technical information related to space. In so doing, it advances knowledge about space, fostering the development and application of space assets by advancing global cooperation. For more information on the IAF, visit *http://www.iafastro.com/*.

International Charter on Space and Major Disasters: Since becoming operational in 2000, the International Charter on Space and Major Disasters has provided a unified system of space data-acquisition and -delivery to those affected by natural or human-caused disasters through authorized users. Each member agency has committed resources to support the provisions of the Charter and thus is helping to mitigate the effects of disasters on human life and property. In order to strengthen the Charter's contribution to disaster management worldwide, member agencies have adopted the principle of universal access to enable non-Charter member countries to submit requests to the Charter for emergency response. For more information on the International Charter on Space and Major Disasters, visit *http://www.disasterscharter.org/web/charter/home*.

International Committee on GNSS (ICG): Created by the United Nations in 2005, the International Committee on Global Navigation Satellite Systems (GNSS) promotes the use of GNSS infrastructure on a global basis and facilitates the exchange of information. The ICG promotes cooperation on civil satellite-based positioning, navigation, timing, and value-added services, as well as compatibility and interoperability among the GNSS systems, while increasing their use to support sustainable development, particularly in the developing countries. The ICG information portal is hosted by the U.N. Office for Outer Space Affairs as a portal for users of GNSS services. For more information on ICG, visit *http://www.unoosa.org/oosa/SAP/gnss/icg.html*.

International Forum for Aviation Research (IFAR): Established in 2011, the International Forum for Aviation Research is a global forum for information exchange and networking by government organizations that are involved in aeronautics-related research. IFAR is open to national aviation research organizations, including universities active in aviation research. IFAR aims to connect the worldwide aviation research community, serve as a venue for information exchange and communication, develop a shared understanding of key challenges faced by the global aviation research community, publish and disseminate information, and identify and evaluate options for new opportunities to cooperate and apply the results of aviation research. For more information on IFAR, visit *http://www.ifar.aero/*.

International Living With a Star (ILWS): Established in 2002, the International Living With a Star program strives to stimulate, strengthen, and coordinate space research to understand the governing processes of the Earth-Sun system as an integrated entity. ILWS encourages the study of the Earth-Sun connected system and its effects on life and society, coordinates collaboration in solar-terrestrial studies and space missions, and provides effective and user-driven access to all data and results from those missions. For more information on ILWS, visit *http://ilwsonline.org/*.

International Mars Exploration Working Group (IMEWG): Established in 1993, the International Mars Exploration Working Group is comprised of representatives from space agencies and major institutions around the world that participate in Mars exploration. Its charter has three aims: to formulate an international strategy for the exploration of Mars, to provide a forum for the coordination of Mars exploration missions, and to examine the possibilities for the next steps beyond currently defined missions. For more information on IMEWG, visit *http://www.bis.gov.uk/ukspaceagency/what-we-do/exploring-the-universe/missions-and-programmes/aurora-space-exploration/global-exploration-strategy/imewg*.

International Ocean-Colour Coordinating Group (IOCCG):
Established by the Committee on Earth Observation Satellites (CEOS) in 1996, the International Ocean-Colour Coordinating Group develops international consensus and synthesis in the subject area of satellite ocean-color radiometry. An international committee of experts heads the group, which is comprised of representatives from national space agencies as well the broader scientific community. The IOCCG fosters expertise in the use of ocean-color data, promotes data-collecting strategies to fill existing time and space, and facilitates the merging of and access to ocean-color data. For more information on IOCCG, visit *http://www.ioccg.org/*.

International Space Education Board (ISEB):
Established in 2005 under the leadership of the education offices of the founding members—the Canadian Space Agency (CSA), European Space Agency (ESA), Japan Aerospace Exploration Agency (JAXA), and NASA—the International Space Education Board has signaled a new era of collaboration in space education. ISEB's mandate is to increase science, technology, engineering, and mathematics literacy achievement in connection with space and to support the future workforce needs of space programs. To that end, ISEB members discuss global issues of

Multilateral Representation and Coordination

importance to each member's outreach and education programs and implements joint education initiatives. For more information on ISEB, visit *http://www.nasa.gov/offices/education/about/iseb-index.html*.

International Space Exploration Coordination Group (ISECG): In 2006, 14 space agencies began a series of discussions on global interests in space exploration that culminated in an articulated vision of peaceful robotic and human space exploration called "The Global Exploration Strategy: The Framework for Coordination." The release of this document in 2007 inspired the establishment of the International Space Exploration Coordination Group, a voluntary, nonbinding international coordination mechanism through which participating agencies exchange information regarding interests, objectives, and plans in space exploration. In 2011, the ISECG released the first iteration of "The Global Exploration Roadmap," which advances the Global Exploration Strategy by articulating the perspectives of participating agencies on exploration goals and objectives, mission scenarios, and coordination of exploration preparatory activities. For more information on ISECG, visit *http://www.globalspaceexploration.org/*.

International Telecommunications Union (ITU): Founded in Paris in 1865 as the International Telegraph Union, the International Telecommunications Union took its present name in 1932, and in 1947 it became a specialized agency of the United Nations for information and communication technologies (ICTs). The ITU allocates the global radio spectrum and satellite orbits, develops the technical standards that ensure that networks and technologies seamlessly interconnect, and strives to improve access to ICTs to underserved communities worldwide. For more information on ITU, visit *http://www.itu.int/en/Pages/default.aspx*.

U.N. Committee on the Peaceful Uses of Outer Space (COPUOS): Established as a permanent committee of the U.N. General Assembly in 1958, the Committee on the Peaceful Uses of Outer Space is charged with promoting the peaceful use of outer space through international cooperation, encouraging the use of information derived from space exploration and research to benefit humankind in general, and helping developing countries build capacity to use space-derived data to mitigate problems on Earth. Growing originally from 18 establishing members, COPUOS now has 71 members, making it one of the largest committees in the U.N. For more information on COPUOS, visit *http://www.oosa.unvienna.org/oosa/COPUOS/copuos.html*.

Index by NASA Mission and Program Area

Aeronautics Research

	Page
Aircraft Icing Research	5, 13, 49
Airframe Noise and Environmental Noise Mitigation	50
Computational Fluid Dynamics (CFD) Research	65
Flight Deck System Design	103
Multilateral Representation and Coordination	139
Sonic Boom Research	96

Education and Outreach

Angry Birds Space	48
Education	128
Exhibits	129
LEGO	26
Multilateral Representation and Coordination	139

Human Exploration and Operations

Astrobiology and Analogs	126
CALorimetric Electron Telescope (CALET)	88
Deep Space Network (DSN)	4, 112
DEvice for the study of CrItical LIquids and Crystallization–Directional Solidification Insert (DECLIC-DSI)	52
Disruption Tolerant Networking (DTN)	89
Global Exploration Strategy (GES)	131
International Astronauts	133
International Space Station (ISS)	15, 39, 94, 107
International Space Station Early Utilization	40
International Space Station Human Research Facility–European Physiology Module (ISS HRF-EPM)	41
International Space Station Utilization: Science and Technology Research and Development	135
Ka-band Radio Frequency (RF) Propagation Monitoring Station	105
Multilateral Representation and Coordination	139
Multi-Purpose Logistics Module (MPLM)	84
NASA Isbjørn Facility at SvalSat	106
Resource Prospector Mission (RPM)	19
Robotic Refueling Mission (RRM)	20

Science

Advanced Colloids Experiment (ACE)	100
Aerosol Robotic Network (AERONET)	124
Alpha Magnetic Spectrometer (AMS)	125
Aquarius/Scientific Application Satellite–D (Aquarius/SAC-D)	2
Astro-H	87

Index by NASA Mission and Program Area

Aura 102, 118
Balloons 127
BepiColombo 79
Cassini-Huygens 29
Chandra X-ray Observatory 64
Cloud-Aerosol LIDAR and Infrared Pathfinder Satellite
Observations (CALIPSO) 51
CloudSat 14
Cluster-II 35
Dawn Asteroid Rendezvous Mission 66, 80
Euclid 36
ExoMars 37
Famine Early Warning System Network (FEWS-NET) 129
Fermi Gamma-ray Space Telescope 53, 67, 81, 90, 114
Global Learning and Observations to Benefit the
Environment (GLOBE) 132
Global Precipitation Measurement (GPM) 91
Global Precipitation Measurement (GPM) Feasibility Study 11
Global Precipitation Measurement (GPM)/Megha-Tropiques 77
Gravity Recovery and Climate Experiment (GRACE) 68
Gravity Recovery and Climate Experiment
Follow-On (GRACE-FO) 69
Hayabusa/Hayabusa-2 92
Heliophysics and Space Physics 101
Herschel 30
Hinode 93, 119
Hubble Space Telescope (HST) 38
Interface Region Imaging Spectrograph (IRIS) 104
Interior Exploration using Seismic Investigations,
Geodesy and Heat Transport (InSight) 54, 70
International Space Weather Initiative (ISWI) 134
Interstellar Boundary Explorer (IBEX) 116
James Webb Space Telescope (JWST) 16, 42
JUpiter ICy moons Explorer (JUICE) 43
Juno 8, 55, 82
Laser Interferometer Space Antenna (LISA) Pathfinder 44
Lunar Reconnaissance Orbiter (LRO) 108
Magnetospheric MultiScale (MMS) 6, 56, 95, 115
Mars Atmosphere and Volatile Evolution (MAVEN) Mission 57
Mars Exploration Rovers (MER) 27, 71
Mars Express 31
Mars Odyssey 58, 109

Index by NASA Mission and Program Area

Mars Reconnaissance Orbiter (MRO)..........................83
Mars Science Laboratory (MSL) 17, 59, 72, 110, 113
Mobile Tracking Station 10
Multilateral Representation and Coordination 139
Nuclear Spectroscopic Telescope Array (NuSTAR).............28, 85
Ocean Surface Topography Mission (OSTM) 60
Oceansat-2..78
Origins Spectral Interpretation Resource Identification Security–
Regolith Explorer (OSIRIS-REx) 18
Ozone Cooperation..12
Planck 32
Rosetta...45
Scientific Application Satellite–C (SAC-C).......................3
SCISAT-1 21
SERVIR..136
Soil Moisture Active Passive Mission (SMAP)...................22
Solar and Heliospheric Observatory (SOHO) 46
Solar Orbiter 33
Solar Probe Plus (SPP)................................9, 61, 73
Solar TErrestrial RElations Observatory (STEREO) 34
Sounding Rockets 137
Space Environment Testbed–1 (SET-1) 62, 120
Space Geodesy...138
Stratospheric Observatory for Infrared Astronomy (SOFIA)........74
Surface Water Ocean Topography (SWOT).....................63
Suzaku...97
Swift Gamma-ray Burst Explorer 86, 121
Terra 23, 98
Time History of Events and Macroscale Interactions during
Substorms (THEMIS) and Acceleration, Reconnection,
Turbulence and Electrodynamics of the Moon's Interaction
with the Sun (ARTEMIS)............................7, 24, 75
Tropical Rainfall Measuring Mission (TRMM)..................99
Two Wide-angle Imaging Neutral-atom Spectrometers (TWINS)....76
Van Allen Probes..25
Vector-Borne Disease Control Research 117
Wind...111
X-ray Multi-Mirror-Newton (XMM-Newton) 47

List of Acronyms

A-Train	Afternoon Constellation
ACE	Advanced Colloids Experiment or Atmospheric Chemistry Experiment
AEB	Brazilian Space Agency (Agência Espacial Brasileira)
AERONET	Aerosol Robotic Network
AFRL	Air Force Research Laboratory (United States)
ALR	Aeronautics and Space Agency (Agentur für Luft- und Raumfahrt) (Austria)
AMS	Alpha Magnetic Spectrometer
APL	Applied Physics Laboratory (United States)
APMS	Aviation Performance Measuring System
APSS	Auxiliary Payload Sensor Subsystem
APXS	Alpha Particle X-ray Spectrometer
ARIS	Active Rack Isolation System
ARTEMIS	Acceleration, Reconnection, Turbulence and Electrodynamics of the Moon's Interaction with the Sun
ASA	Austrian Space Agency
ASI	Italian Space Agency (Agenzia Spaziale Italiana)
ASINet	ASI Network Infrastructure
ASPERA-3	Analyzer of Space Plasmas and Energetic Atoms Version 3
ASPOC	Active Spacecraft Potential Control
ASTEP	Astrobiology Science and Technology for Exploring Planets
ASTER	Advanced Spaceborne Thermal Emission and Reflection Radiometer
ASTID	Astrobiology Science and Technology Instrument Development
ATV	Automated Transfer Vehicle
BAO	Baryonic Acoustic Oscillations
BARREL	Balloon Array for Radiation belt Relativistic Electron Losses
BELSPO	Belgian Federal Science Policy Office
BLAST	Balloon-borne Large Aperture Sub-millimeter Telescope
CAB	Center for Astrobiology (Spain)
CALET	CALorimetric Electron Telescope
CALIPSO	Cloud-Aerosol LIDAR and Infrared Pathfinder Satellite Observations
CATHALAC	Water Center for the Humid Tropics of Latin America and the Caribbean
CCAD	Central American Commission for Environment and Development

List of Acronyms

CCAFS	Cape Canaveral Air Force Station (United States)
CCD	Charge-Coupled Device
CCF	Capillary Channel Flow
CCSDS	Consultative Committee for Space Data Systems
CDDIS	Crustal Dynamics Data Information System
CDSCC	Canberra Deep Space Communication Complex (Australia)
CDTI	Center for the Development of Industrial Technology (Spain)
CEA	Alternative Energies and Atomic Energy Commission (Commissariat à l'énergie atomique et aux énergies alternatives) (France)
CEA Saclay	CEA Research Center at Saclay (France)
CEM	Correlative Environment Monitor
CEOS	Committee on Earth Observation Satellites
CERN	European Organization for Nuclear Research (Conseil Européen pour la Recherche Nucléaire)
CFD	Computational Fluid Dynamics
CFE	Capillary Flow Experiment
CGBA	Commercial Generic Bioprocessing Apparatus
CHAMPS	Charge and Mass of Meteoric Smoke Particles
CHD	Charge Detector
ChemCam	Laser-Induced Remote Sensing for Chemistry and Micro-Imaging
CIR	Co-rotating Interaction Region
CMB	Cosmic Microwave Background
CME	Coronal Mass Ejection
CNES	National Center for Space Studies (Centre National d'Études Spatiales) (France)
CNU	Chungnam National University (Republic of Korea)
CONAE	National Commission on Space Activities (Comisión Nacional de Actividades Espaciales) (Argentina)
COPUOS	Committee on the Peaceful Uses of Outer Space (United Nations)
COSPAR	Committee on Space Research
COTS-2	Commercial Off-The-Shelf–2
CREDANCE	Cosmic Radiation Environment Dosimetry and Charging Experiment
CRS	Commercial Resupply Services
CSA	Canadian Space Agency
CSIRO Research	Commonwealth Scientific and Industrial Organization (Australia)

CSL	Liège Space Center (Centre Spatial de Liège) (Belgium)
DAN	Dynamic Albedo of Neutrons
DECLIC	DEvice for the study of Critical LIquids and Crystallization
Dextre	Special Purpose Dexterous Manipulator
DFM	Digital Fluxgate Magnetometer
DIS	Dual Ion Spectrometers
DLN	Digital Learning Network
DLR	German Aerospace Center (Deutsches Zentrum für Luft- und Raumfahrt)
DORIS	Doppler Orbitography and Radiopositioning Integrated by Satellite
DPR	Dual-frequency Precipitation Radar
DRS	Disturbance Reduction System
DRTS	Data Relay Test Satellite
DSI	Directional Solidification Insert
DSN	Deep Space Network
DSS	Deep Space Station
DSX	Demonstration and Science Experiments
DTN	Disruption Tolerant Networking
EarthKAM	Earth Knowledge Acquired by Middle School Students
EDI	Electron Drift Instrument
EDL	Entry, Descent, and Landing
EDM	EDL Demonstrator Module
EF	Exposed Facility
EIK	Extended Interaction Klystrons
EIS	Extreme-ultraviolet Imaging Spectrometer
EIT	Extreme-ultraviolet Imaging Telescope
ELENA	Emitted Low-Energy Neutral Atoms
ELM	Experiment Logistics Module
ELV	Expendable Launch Vehicle
EMCS	European Modular Cultivation System
EMFISIS	Electric and Magnetic Field Instrument Suite and Integrated Science
ENA	Energetic Neutral Atom
EO	Earth Observing
EOS	Earth Observing System
EPHIN	Electron, Proton, Helium Instrument
EPIC	European Photon Imaging Camera
EPM	European Physiology Module
ERA	European Robotic Arm

List of Acronyms

ESA	European Space Agency
ESTEC	European Space Research and Technology Centre
EUMETSAT	European Organisation for the Exploitation of Meteorological Satellites
EVA	Extravehicular Activities
EXPRESS	EXpedite the PRocessing of Experiments to Space Station
FC	Framing Cameras
FCB	Functional Cargo Block
FEWS-NET	Famine Early Warning System Network
FFG	Austrian Research Promotion Agency (Die Österreichische Forschungsförderungsgesellschaft) (Austria)
FGM	Fluxgate Magnetometer
FGS	Fine Guidance Sensor
FIR	Fluids Integration Rack
FMI	Finnish Meteorological Institute
FOC	Faint Object Camera
FPI	Fast Plasma Instrument
FTS	Fourier Transform Spectrometer
GBM	GLAST Burst Monitor
GBO	Ground-Based Observatory
GDE	Gun Detector Electronics
GEO	Group on Earth Observation
GEOSS	Global Earth Observation System of Systems
GES	Global Exploration Strategy
GFZ	German Research Centre for Geosciences (Deutsches GeoForschungsZentrum)
GGOS	Global Geodetic Observing System
GGS	Global Geospace Science
GLAST	Gamma-ray Large Area Space Telescope
GLOBE	Global Learning and Observations to Benefit the Environment
GMI	GPM Microwave Imager
GNSS	Global Navigation Satellite Systems
GOLPE	GPS Occultation and Passive reflection Experiment
GPM	Global Precipitation Measurement
GPS	Global Positioning System
GRACE	Gravity Recovery and Climate Experiment
GRACE-FO	Gravity Recovery and Climate Experiment Follow-On
GRAIL	Gravity Recovery and Interior Laboratory

GRB	Gamma-Ray Burst
GRC	Glenn Research Center (United States)
GRS	Gamma Ray Spectrometer
GSFC	Goddard Space Flight Center (United States)
GSOC	German Space Operations Center
HEND	High-Energy Neutron Detector
HFI	High-Frequency Instrument
HGA	High-Gain Antenna
HIFI	Heterodyne Instrument for the Far-Infrared
HIRDLS	High-Resolution Dynamic Limb Sounder
HIS	Heavy Ion Spectrometer
HIWC	High Ice Water Content
HP3	Heat Flow and Physical Properties Package
HRC	High-Resolution Camera
HRF	Human Research Facility
HRPC	High-Resolution Panchromatic Camera
HRSC	High-Resolution Stereo Camera
HSO	Hungarian Space Office
HST	Hubble Space Telescope
HSTC	High-Sensitivity Technological Camera
HTV	H-II Transfer Vehicle
HXD	Hard X-Ray Detector
HXI	Hard X-Ray Imager
IAC	International Astronautical Congress
IADC	Interagency Space Debris Coordination Committee
IAF	International Astronautical Federation
IAG	International Association of Geodesy
IAP	Institute of Atmospheric Physics (Czech Republic)
IBEX	Interstellar Boundary Explorer
ICARE	Influence of Space Radiation on Advanced Components
ICG	International Committee on GNSS
ICRR	Institute for Cosmic-Ray Research (Japan)
ICT	Information and Communication Technology
IDS	Instrument Deployment System
IES	Ion and Electron Sensor
IFAR	International Forum for Aviation Research
IGS	International GPS Service
IIR	Imaging Infrared Radiometer
IKI RAS	Space Research Institute of the Russian Academy of Sciences

List of Acronyms

ILRS	International Laser Ranging Service
ILWS	International Living With a Star
IMC	Imaging Calorimeter
IMEWG	International Mars Exploration Working Group
IMPACT	In-situ Measurements of Particles and CME Transients
IN2P3	National Institute of Nuclear and Particle Physics (Institut national de physique nucléaire et de physique des particules) (France)
INES	Italian Navigation Experiment
INPE	National Institute for Space Research (Instituto de Pesquisas Espaciais) (Brazil)
InSight	Interior Exploration using Seismic Investigations, Geodesy and Heat Transport
INTA	National Institute of Aerospace Technology (Instituto Nacional de Técnica Aeroespacial) (Spain)
IOAG	Interagency Operations Advisory Group
IOCCG	International Ocean-Colour Coordinating Group
IRF	Institute of Space Physics (Institutet för rymdfysik) (Sweden)
IRIS	Interface Region Imaging Spectrograph
IRT	Icing Research Tunnel (United States)
ISAS	Institute of Space and Astronautical Science (Japan)
ISEB	International Space Education Board
ISECG	International Space Exploration Coordination Group
ISERV	ISS SERVIR Environmental Research and Visualization
ISIM	Integrated Science Instrument Module
ISPR	International Standard Payload Rack
ISRO	Indian Space Research Organisation
ISS	International Space Station
IST	Italian Star Tracker
ISTP	International Solar-Terrestrial Physics
ISWI	International Space Weather Initiative
ITU	International Telecommunications Union
IUVS	Imaging Ultraviolet Spectrograph
IVS	International VLBI Service
IWF/ÖAW	Space Research Institute of the Austrian Academy of Science (Institut für Weltraumforschung, Österreichischen Akademie der Wissenschaften) (Austria)
JADE	Jovian Auroral Distributions Experiment

List of Acronyms

JAXA	Japan Aerospace Exploration Agency
JEM	Japanese Experiment Module
JIRAM	Jovian Infrared Auroral Mapper
JPL	Jet Propulsion Laboratory (United States)
JSC	Johnson Space Center (United States)
JUICE	JUpiter ICy moons Explorer
JWST	James Webb Space Telescope
KaRIn	Ka-band Radar Interferometer
KASI	Korea Astronomy and Space Science Institute
KaT	Ka-band Transponder
KSAT	Kongsberg Satellite Services (Norway)
KSC	Kennedy Space Center (United States)
KTH	Royal Institute of Technology (Kungliga Tekniska Högskolan) (Sweden)
LANCE	Land Atmosphere Near-real-time Capability for EOS
LASCO	Large Angle and Spectrometric Coronograph Experiment
LAT	Large Area Telescope
LCROSS	Lunar Crater Observation and Sending Satellite
LEND	Lunar Exploration Neutron Detector
LEO	Low-Earth Orbit
LETGS	Low-Energy Transmission Grating for Cosmic X-ray Spectrometer
LFI	Low-Frequency Instrument
LIDAR	Light Detection and Ranging
LISA	Laser Interferometer Space Antenna
LLNL	Lawrence Livermore National Laboratory (United States)
LMM	Light Microscopy Module
LPP	Laboratory of Plasma Physics (France)
LPW	Langmuir Probe and Waves
LRA	Laser Retroreflector Array
LRI	Laser Ranging Interferometer
LRO	Lunar Reconnaissance Orbiter
LTP	LISA Technology Package
LVPS	Low-Voltage Power Supply
LWS	Living With a Star
MADRAS	Microwave Analysis and Detection of Rain and Atmospheric Structures
MAESTRO	Measurements of Aerosol Extinction in the Stratosphere and Troposphere Retrieved by Occultation

List of Acronyms

MaRS	Mars Radio Science Experiment
MARSIS	Mars Advanced Radar for Subsurface and Ionospheric Sounding
MAVEN	Mars Atmosphere and Volatile Evolution
MB	Mössbauer Spectrometer
MCOP	Multilateral Crew Operations Panel
MDSCC	Madrid Deep Space Communications Complex (Spain)
MEEMM	Multi-Electrode Electroencephalogram Mapping Module
MELFI	Minus-80°C Laboratory Freezer for ISS
MER	Mars Exploration Rover
METI	Ministry of Economy, Trade and Industry (Japan)
MIPA	Miniature Ion Precipitation Analyzer
MIRI	Mid-Infrared Instrument
MIRO	Microwave Instrument for the Rosetta Orbiter
MISR	Multiangle Imaging SpectroRadiometer
MLM	Multipurpose Laboratory Module
MLS	Microwave Limb Sounder
MMO	Mercury Magnetospheric Orbiter
MMP	Magnetic Mapping Payload
MMRS	Multispectral Medium Resolution Scanner
MMS	Magnetospheric MultiScale
MODIS	Moderate Resolution Imaging Spectroradiometer
MOMA	Mars Organic Molecule Analyzer
MoonKAM	Moon Knowledge Acquired by Middle School Students
MOPITT	Measurements of Pollution in the Troposphere
MPE	Max Planck Institute for Extraterrestrial Physics (Germany)
MPLM	Multi-Purpose Logistics Modules
MPO	Mercury Planetary Orbiter
MR	Microwave Radiometer
MRI	Microwave Ranging Instrument
MRO	Mars Reconnaissance Orbiter
MSFC	Marshall Space Flight Center (United States)
MSG	Microgravity Science Glovebox
MSL	Mars Science Laboratory
MSRR-1	Materials Science Research Rack-1
MSS	Mobile Servicing System
NAI	NASA Astrobiology Institute
NAOJ	National Astronomical Observatory of Japan

NBI	Niels Bohr Institute for Astronomy, Physics, and Geophysics (Denmark)
NDVI	Normalized Difference Vegetation Index
NEN	Near Earth Network
NERC	National Environment Research Council (United Kingdom)
NGIMS	Neutral Gas and Ion Mass Spectrometer
NGSLR	Next-Generation Satellite Laser Ranging
NHSC	NASA Herschel Science Center
NICT	National Institute of Information and Communications Technology (Japan)
NIR	Near-Infrared
NIRCam	Near-Infrared Camera
NIRISS	Near-Infrared Imager and Slitless Spectrograph
NIRSpec	Near-Infrared Spectrograph
NIRST	New Infrared Sensor Technology
NISP	Near-Infrared Spectrograph and Photometer
NLR	Netherlands National Aerospace Laboratory (Nationaal Lucht- en Ruimtevaartlaboratorium)
NOAA	National Oceanic and Atmospheric Administration (Unites States)
NRC	National Research Council (Canada)
NSC	Norwegian Space Centre
NSRP	NASA Sounding Rocket Program
NuSTAR	Nuclear Spectroscopic Telescope Array
ÖAW	Austrian Academy of Science (Österreichische Akademie der Wissenschaften)
OBA	Optical Bench Assembly
OCAMS	OSIRIS-REx Camera Suite
OCM	Ocean Color Monitor
OLA	OSIRIS-REx Laser Altimeter
OM	Optical/Ultraviolet Monitor
OMEGA	Visible and Infrared Mineralogical Mapping Spectrometer
OMI	Ozone Monitoring Instrument
ONERA	The French Aerospace Lab (Office National d'Études et Recherches Aérospatiales) (France)
OSIRIS-REx	Origins Spectral Interpretation Resource Identification Security–Regolith Explorer
OSTM	Ocean Surface Topography Mission
OTE	Optical Telescope Element
OTES	OSIRIS-REx Thermal Emission Spectrometer
OVIRS	OSIRIS-REx Visible and Infrared Spectrometer
PACS	Photoconductor Array Camera and Spectrometer

List of Acronyms

PARASOL	Polarization and Anisotropy of Réflectances for Atmospheric Sciences coupled with Observations from a LIDAR
PEP	Particle Environment Package
PFS	Planetary Fourier Spectrometer
PICAM	Planetary Ion Camera
PIV	Particle Image Velocimetry
PLASTIC	PLAsma and SupraThermal Ion Composition
PM	Pressurized Module
PMM	Permanent Multipurpose Module
RAD	Radiation Assessment Detector
REMS	Rover Environmental Monitoring Station
REXIS	Regolith X-ray Imaging Spectrometer
RF	Radio Frequency
RFU	Radio Frequency Unit
RGS	Reflection Grating Spectrometer
RIME	Radar for Icy Moon Exploration
RISE	Rotation and Interior Structure Experiment
RMS	Remote Manipulator System
ROSA	Radio Occultation Sounder for Atmosphere
Roscosmos	Russian Federal Space Agency
ROSINA	Rosetta Orbiter Spectrometer for Ion and Neutral Analysis
RPC	Rosetta Plasma Consortium
RPM	Resource Prospector Mission
RRM	Robotic Refueling Mission
RRS	Research Range Services
SAC-C	Scientific Application Satellite–C
SAC-D	Scientific Application Satellite–D
SAM	Sample Analysis at Mars
SAPHIR	Sounder for Probing Vertical Profiles of Humidity
SAR	Synthetic Aperture Radar
ScaRaB	Scanner for Radiation Budget
SCM	Search Coil Magnetometer
SDO	Solar Dynamics Observatory
SDP	Spin-Plane Double Probe
SECCHI	Sun Earth Connection Coronal and Heliospheric Investigation
SEIS	Seismic Experiment for Interior Structure
SERENA	Search for Exospheric Refilling and Emitted Natural Abundances
SET-1	Space Environment Testbed–1
SGD	Soft Gamma-ray Detector

List of Acronyms

SGR	Soft Gamma Repeater
SHADOZ	Southern Hemisphere ADditional OZonsondes
SHARAD	Shallow Subsurface Radar
SLD	Supercooled Large Droplet
SLR	Satellite Laser Ranging
SMA	Scan Mirror Assembly
SMAP	Soil Moisture Active Passive
SNSB	Swedish National Space Board
SOFIA	Stratospheric Observatory for Infrared Astronomy
SOHO	Solar and Heliospheric Observatory
SoloHI	Solar Orbiter Heliospheric Imager
SOT	Solar Optical Telescope
SPICAM	Spectroscopy for Investigation of Characteristics of the Atmosphere of Mars
SPIRE	Spectral and Photometric Imaging Receiver
SPP	Solar Probe Plus
SSO	Swiss Space Office
SSRMS	Space Station Remote Manipulation System
SST	Sea Surface Temperature
STATIC	SupraThermal and Thermal Ion Composition
STEM	Science, Technology, Engineering, and Mathematics
STEREO	Solar TErrestrial RElations Observatory
STS	Space Transportation System
SvalSat	Svalbard Satellite Station (Norway)
SWA	Solar Wind Plasma Analyzer
S/WAVES	STEREO/Wind/Radio and Plasma Wave Experiment
SWEA	Solar Wind Electron Analyzer
SWEAP	Solar Wind Electrons Alphas and Protons
SWIA	Solar Wind Ion Analyzer
SWOT	Surface Water Ocean Topography
SXI	Soft X-ray Imager
SXS	Soft X-ray Spectrometer
TAGSAM	Touch-and-Go Sample Mechanism
TASC	Total AbSorption Calorimeter
Tekes	Finnish Funding Agency for Technology and Innovation (Tekniikan Edistämiskeskus)
TES	Tropospheric Emission Spectrometer
THEMIS	Time History of Events and Macroscale Interactions during Substorms
TKSC	Tsukuba Space Center
TMI	TRMM Microwave Imager

List of Acronyms

TNR	Thermal Noise Receiver
TNSC	Tanegashima Space Center (Japan)
TRMM	Tropical Rainfall Measuring Mission
TT&C	Telemetry, Tracking, and Command
TWINS	Two Wide-angle Imaging Neutral-atom Spectrometers
USAID	U.S. Agency for International Development
USGS	U.S. Geological Survey
UVOT	Ultraviolet/Optical Telescope
UVS	Ultraviolet Spectrometer
VAFB	Vandenberg Air Force Base (United States)
VIR-MS	Visual and Infrared Mapping Spectrometer
VIS	Visual Imager
VLBI	Very Long Baseline Interferometry
WAVES	Wind/Radio and Plasma Wave Experiment
WFC	Wide-Field Camera
WFF	Wallops Flight Facility (United States)
WISPR	Wide-field Imager for Solar Probe
WL	Weak Gravitational Lensing
XIS	X-Ray Imaging Spectrometer
XMM	X-Ray Multi-Mirror
XRS	X-Ray Spectrometer
XRT	X-Ray Telescope